RONALD RUSSELL

Discovering
Lost Canals

SHIRE PUBLICATIONS LTD.

Key to symbols used on the maps in this book

▬ ▬ ▬	**Lost Canal**	┼┼┼┼	**Existing Canal**
Ⓐ	**Aqueduct**	┼─┼─┼	**Railway**
ⒾⓅ	**Inclined Plane**	〜〜	**River**
Ⓛ	**Lift**	▬▬▬	**Main Road**

ACKNOWLEDGEMENTS

The author is grateful for help received from officials of various canal trusts and societies while preparing this second edition. Cartography by Richard G. Holmes. Cover design by Ron Shaddock.

Printed in Great Britain by Hunt Barnard Printing Ltd, Aylesbury, Bucks.

Contents

Introduction

Canals provided the chief method of inland transport in Britain for some eighty years. In 1765 the Duke of Bridgewater's canal, engineered by James Brindley, was opened from his mines at Worsley. By this time the St Helen's Canal (or Sankey Brook Navigation) had already been operating for eight years, but it was the Duke's canal that provided the impetus. Soon work began on the Staffordshire & Worcestershire, the Trent & Mersey, the Coventry, Birmingham and Oxford canals, among many others. A host more were begun in the 1790s, including the Monmouthshire, Grand Junction, Shrewsbury, Warwick & Birmingham, Rochdale, Kennet & Avon and Wilts & Berks. On most waterways coal was the main traffic and its price in towns and villages fell dramatically. Culm, lime, timber, salt, building materials, agricultural produce and general merchandise were also distributed widely all over the country and centres of population served by canals developed at a remarkable rate.

With the accession of Victoria, however, came the beginning of the railway age. In 1846 nearly six hundred miles of canal were sold or leased to railway companies. Slow and unglamorous, most canals proved easy victims; toll-cutting helped to retain traffic for a time but revenue fell and the standard of maintenance deteriorated. The decline continued steadily; by the advent of the twentieth century few canals were trading profitably and the coming of motorised road transport added to their difficulties. By the Second World War, very few, notably the Aire & Calder and the consortium of canals known as the Grand Union, were really in business.

After the war most of the surviving canals were nationalised, but this at first did not stop the closures. Then, with the formation of the Inland Waterways Association, a voluntary body, a change of attitude gradually became apparent. People became aware that this network of waterways covering most of the country could provide wonderful opportunities, if not for commercial transport, then for recreation and leisure. Pleasure cruising became a popular activity, and attention turned to some of the disused waterways to see if they could be brought back to life. In 1964 the Stratford-on-Avon Canal was reopened and since then restoration work has been proceeding across the country, from Pocklington in Humberside to Tiverton in Devon. On a handful of waterways, mostly in the North-east, some trade has continued; now there are ideas for bringing back more commercial traffic to the canals, and the Inland Shipping Group has been formed to investigate the potential and to put forward schemes for new waterways to relieve our over-congested roads. The British Waterways Board, although handicapped by lack of finance, are enthusiastic about the future

4

of the system and there is some hope that, if government and local authorities give full and tangible support, a new Canal Age will dawn.

At the present time, however, there are about a hundred abandoned and derelict canals to be found in England and Wales. For some of them there is hope of resuscitation, in part if not in whole: the Thames & Severn and the Wey & Arun Junction are two examples. But most of them have decayed too far, or have had much of their course obliterated by railway or building development. Yet there is still much to be found, especially on the lines of those which still run through agricultural country. Bridges and lock remains abound; there are tunnels, aqueducts, cuttings ·and embankments, warehouses and the sites of wharves. There are engineering and architectural curiosities, like the sites of inclined planes, the Grand Western lifts, the guillotine gates on the Shrewsbury Canal and the round houses on the Thames & Severn. Exploring them takes you to parts of the country you might otherwise never have visited. The only equipment you need is the appropriate Ordnance Survey map: the 1:50,000 scale map will do, but the 2½ or 6-inch are better. Stout footwear is recommended, and a compass or a good sense of direction will help. You should also remember to ask permission before you venture on to private land.

In this book you will find brief historical· accounts and itineraries of sixteen 'lost' canals, including the Shropshire tub-boat system. There is also a list of several more which the reader may care to explore for himself. For further information, there is a list of books, some of which deal with individual waterways and give detailed accounts of their remains. County record offices and public libraries can also help if you want to find out more.

Finally I would like to pay tribute to the work of the pioneer canal historian, Charles Hadfield, on whose researches I have largely drawn for the historical background of the various canals, and to the many people, especially Adrian Russell, who helped me with the fieldwork on which this book is based.

1. The Bude Canal

Act: 1819. Opened: 1823. Abandoned: 1891 (except Bude-Marhamchurch, abandoned 1960). Length: 35½ miles.

History

Some of the most ambitious canal schemes originated in the south-west of England, in counties where the countryside provided very testing problems for the construction of artificial waterways. In 1774 a canal of more than 90 miles was proposed from Bude by a serpentine route to Calstock on the Tamar. John Edyvean, who put forward the idea, was himself the promoter of another Cornish canal, the St Columb; like the St Columb (of which traces can be found near St Columb Major and St Columb Porth, although it was never finished) the Bude was intended mainly to carry sea sand to the farms inland for manure. It was likewise to be used by tub-boats and the gradients were to be dealt with by inclined planes. An Act was obtained in 1774, but the financial outlook was not very attractive and nothing was done. Interest in the idea revived some years later and James Green was asked to produce a plan. After some modifications to his scheme, another Act was obtained in 1819 authorising certain improvements to the harbour at Bude and the making of a canal from Bude to Thornbury with branches to Druxton, near Launceston, and to a reservoir later known as Tamar Lake. The total length was to be nearly 46 miles. From Bude to Helebridge the canal would take barges; the remainder would be narrow for tub-boats 20 feet long with a beam of 5 feet 6 inches. The estimated cost was £91,617.

Most of the work was completed by July 1823 when the canal was opened from Bude to Blagdonmoor, near Holsworthy (which became the terminus instead of Thornbury), to Tamar Lake and to Tamerton bridge. The line from Tamerton bridge to Druxton was finished two years later. The total cost of the undertaking was about £118,000, and 35½ miles in all were cut. The cost included improvements to the harbour, a sea lock and two locks on the broad stretch, and six inclined planes. There was a rise of 433 feet to the summit on the Holsworthy line.

The Bude tub-boats, which took about 5 tons, operated in gangs of up to six, drawn by a horse. Like fen lighters, they were steered from the second boat in line. They were distinguished in that they were equipped with wheels which fitted into the rails on the inclines. The inclines were double-track, all except the longest, Hobbacott, operated by a water-wheel. Boats were attached to a chain and drawn up or eased down in a matter of minutes. The Hobbacott incline worked on the bucket-in-a-well system, like the Wellisford incline on the Grand Western was intended to do. Here there was a steam engine kept in reserve for times when the chain lowering the vast bucket holding 15 tons of water snapped. Spare buckets were also kept at the top. Sometimes the chains holding the boats broke. Life at Hobbacott was often exciting but also expensive to the canal company.

Sand was the main cargo carried; coal, culm, limestone, slate and building materials were also distributed. Traffic, however, never came up to expectations and the annual tonnage seldom exceeded 50,000. In some years small profits were made but not until 1876 was the company able to pay its first dividend. By this time railways were serving the area and other fertilisers than sand were available to the farmers. Traders began to desert the canal but it managed to struggle on through the 1880s. The company decided on abandonment of the navigation; this ended in 1891 apart from the barge canal which continued for a few more years. In 1901 the Stratton & Bude UDC bought the line from Bude to the reservoir for water supply. Much of it was sold off in the 1960s by the Council (now called the Bude-Stratton UDC); the reservoir and the length of the canal leading to it went to the North Devon Water Board and various parts of the line were bought by farmers. Only the bottom two miles at Bude are now owned by the Council, apart from a stretch at Anderton which no-one has bought.

The canal today

It is not particularly difficult to find the Bude Canal but most of its course is now in private ownership and permission to explore must be obtained. If you have a car, however, it is possible to visit most of the more interesting remains within a day, provided you have a copy of the OS sheet with you.

Bude, a pleasant seaside town, is still a registered port. There is a massive breakwater, built to replace the one constructed under the 1819 Act which was destroyed during a storm in 1838. There is the impressive sea lock that could take vessels of up to 300 tons. There are some early canal buildings by the lock, and on the north side traces of a narrow railway which was made so that sand could be pulled up in trucks for loading into barges and tub-boats. You can walk alongside the canal to see the two locks, Roddsbridge and Whalesborough.

The A39 crosses the canal at **Helebridge,** two miles south of Bude. The basin is on the east side; here barges transferred their cargoes to the tub-boats. A minor road leads east to Marhamchurch; parallel to this on its north side behind a hedge is the **Marhamchurch inclined plane,** 836 feet long, rising 120 feet. You may be able to scramble up the path alongside the slope. From the top of the incline the canal course winds north-eastwards across the fields; some of it has been ploughed in. Make your way to the A3072 Stratton-Holsworthy road. Half-way between Stratton and the Red Post cross-roads is Thurlibeer Farm, with

7

the **Hobbacott Down inclined plane** in its grounds. At the top there are some old buildings and the entrance to one of the bucket wells. The line of the incline — 935 feet long with a 225-foot rise — is clear, and there is a splendid view from the top. Continue to **Red Post;** a short distance south on B3254 the road once crossed the canal by a bridge and the southern branch to Druxton swung off a few yards to the east.

To follow the main line, resume the road towards Holsworthy and take the first left. There is a bridge on this road; you may be able to follow the course to the stone-built **Burmsdon aqueduct** across the Tamar. On the east of this is the **Venn, or Veala, incline,** with a 58-foot rise. Water supply is now piped along this route from **Tamar Lake.** You can reach this point by taking the second left past Red Post, where there is a lane off to the left just over a mile along. This road reaches a T junction; the northern arm to Tamar Lake, which fed the canal, turns off a couple of hundred yards north-west of the junction. This feeder is marked on the OS map and can be followed northwards through Puckland to the lake.

For the main line, having looked for an embankment near the road, turn right to rejoin the Holsworthy road. The canal meanders through fields on the north side; it is discernible from most of the left turns towards **Holsworthy,** which describes itself as a Port Town and where Stanbury Wharf, now overgrown but with a warehouse and wharfinger's cottage still standing, was a quite important trading centre. In Holsworthy, take A388 for Bideford and turn left after half a mile. This road crosses the canal twice in just over a mile. Return to A388, turn towards Holsworthy and take the first left. You are now travelling parallel to the last stretch of the main line. In 1½ miles a lane on the left leads to **Blagdonmoor wharf,** where there are a number of buildings, warehouses and cottages. The line of the canal continues through a cutting, where a tunnel was to be built for the extension to Thornbury.

To explore the southern branch, return to Red Post and turn left. The line was cut roughly parallel to the Tamar and close to it. There are three inclines on this arm. The first is **Merrifield,** falling 60 feet. This lies to the north-east of the derelict Whitstone & Bridgerule station, three miles south of Red Post and to the left of the road. For the next, **Tamerton,** continue on B3254 for another four miles and turn left for North Tamerton. About a quarter of a mile east of the church on this, the Holsworthy road, was a low aqueduct which impeded the passage of traffic, especially steam traction engines. This was demolished after the canal closed, but the abutments can be seen. On the north side of the road was a wharf and on the south a coal store; some of the buildings have been converted into bungalows. Take the road heading south out

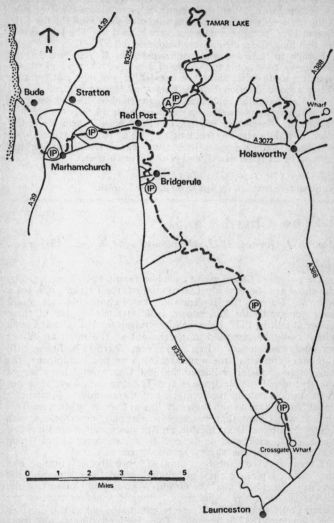

The Bude Canal

9

of **North Tamerton;** in about half a mile you come to a farm at Tamerton Town. Then walk south-eastwards along a lane and paths for a mile to reach the incline, ploughed over for the first time in 1978, when several relics were turned up. The incline was 360 feet long, the same as Merrifield, and fell 59 feet.

Return to the road and continue south. If you turn left at **Boyton** you soon come to the site of a bridge, a basin and wharf. Walking southwards along the canal line, you will reach a surviving accommodation bridge, known as Hunch bridge.

For the last incline, continue along the road to Bridgetown; turn left at the T junction and soon you pass under a bridge — which is part of the **Werrington incline.** The head is in a farm above the road. This slope was 259 feet long and fell 51 feet. Now turn about and head for **Crossgate;** here some canal buildings line the remains of Druxton wharf, and your journey is over.

OS sheet 190 (1:50,000), 174 (1 inch).
Bude can be reached by bus from Exeter Station.

2. The Chard Canal

Act: 1834. Opened: 1842. Abandoned: 1867. Length: 13½ miles.

History

The people of Chard might have had reason to feel neglected by the canal age. Several ideas were mooted, the first in 1769, for a waterway between the English and Bristol channels which might have penetrated the Blackdown Hills at Chard. None of these materialised. In 1827, however, the Bridgwater & Taunton Canal, which could be regarded as a remnant of the more ambitious schemes, was opened. A first move from Chard to be linked with this canal met with little response, but a proposal to improve the navigation of the river Parrett and link Chard with that changed the attitude of the Bridgwater and Taunton company. They put their support behind the formation of a company to construct a canal from Chard to join their own line at Creech St Michael. This time it was Chard that was slow to respond; nevertheless, with energetic support from Bristol, an Act was obtained and most of the £57,000, estimated as the cost by the engineer James Green who carried out the survey, was soon raised.

Green proposed a tub-boat canal with two lifts, two tunnels and two inclined planes; the line, though short, was through hilly country. Green's lifts on the Grand Western had caused trouble; he was succeeded as engineer by Sydney Hall before work on the Chard Canal began. The lifts were eliminated; when the line was completed in 1842 it included four inclined planes and three tunnels, as well as two aqueducts and two locks. The cost had risen to about £140,000.

The hopes of the promoters came nowhere near to realisation. Although there was a reasonable amount of traffic in the first few years, mainly coal to Chard, the company could not pay off its mortgages and was always in financial difficulty. One can now see that the canal was opened far too late to have any prospect of success, for the railway era had begun and the Bristol & Exeter line was opened to Taunton in the same year that trading began on the canal. Indeed, in 1846 the company wanted to convert part of the canal into a railway but were not permitted to do so until the mortgages had been settled. So the canal continued into the 1860s, carrying between 20,000 and 30,000 tons a year. Then the Bristol & Exeter Railway, fighting off competition from the London & South Western, bought both the Bridgwater & Taunton and the Chard canals. The price paid for the Chard was £5,945. The canal was closed under an Act of 1867 and the reservoir, incline machinery and parts of the line were sold.

The canal today

There is enough left of the Chard Canal to make a good day's exploration. If you start from Chard, try to obtain a copy of the Chard History Group's booklet on the canal, which provides a most useful guide. **Chard basin** was on the north side of the town; look for the buildings of B. G. Wyatt which include two canal warehouses and part of the basin boundary wall. The main road to follow out of Chard is A358; for the whole of its length the canal lies to the east of this road, never more than two miles from it.

There may be a few traces of the canal between Wyatt's premises and the reservoir, but much of it has been obliterated. The road to Chaffcombe, leaving A358 a mile north of Chard, leads to the reservoir. Opposite the northern end of the reservoir a footpath can be followed to the site of the **Chard Common inclined plane.** This had a fall of 86 feet and, unlike the other Chard inclines, was only single track, boats being carried in a cradle and the power provided by water turbine.

The road from A358 to Knowle St Giles has a good stretch of canal bed on its northern side, just to the east of where the road crosses the track of the old railway. It heads north-eastward, with the remains of a lock north of a covert, about a mile from the road. If you take the A3037 to Ilminster and stop by the junction with the road to Kingstone you will see some cottages built on the canal bed. The canal has now turned to head north-north-west past Ilminster. Much of this stretch has recently disappeared beneath development, but you should be able to find the preserved north portal of **Ilminster tunnel,** which was 300 yards long. From this portal an incline took the level down 82½ feet; the foot of the incline was by the boundary of the school playing field.

The canal is crossed by A303 half a mile west of the centre of Ilminster. The aqueduct over the Isle to the north was later used by the railway. There is nothing much remaining in the next three miles. Make for the village of **Beercrocombe,** to which there are several turnings off A358. There are remains of an embankment south and east of the village. Close to the road on the north of the village was the entrance to the canal's longest tunnel, **Crimson Hill.** This portal has collapsed, but the northern portal survives near Wrantage. **Wrantage** is on A378, 1¾ miles east of the junction with A358. The Canal Inn is a useful landmark. You can see the abutment of an aqueduct here. Follow the line south-eastward; there is a good stretch of canal, with a wide pound at the foot of the Wrantage incline. There is a rise of 27½ feet. Near to the top is the tunnel portal. This tunnel was 1,800 yards long; in the roof there are shackles which aided the boatmen in hauling the boats through. You will also notice holes in the tunnel side, probably for drainage. A pump now extracts water from the tunnel for farm use. There was a tunnel keeper's house of which the remains may still exist.

From Wrantage head west for half a mile along A378 and turn right for Lillesdon. This road soon crosses the canal, which continues north-west in a cutting leading to the south portal of **Lillesdon tunnel,** of which recently only the top of the arch was visible. This tunnel is 314 yards long; if you continue towards Lillesdon and turn left to return to A378 you cross the line of the tunnel. The north portal is in a field to the north of the road and can be detected by a sharp fall in the gradient.

Now continue to A358 and turn off for Thornfalcon. The road leading north out of the village passes by Canal Farm, where it may be possible to find the site of the fourth incline. The **Thornfalcon incline** took the level down another 28 feet. Keep on this road towards Creech, but turn left at a T junction for **Ruishton.** Soon you will come to the abutments of an aqueduct over the road. From here the canal was carried on an impressive embankment across the flood plains of the river Tone and then across the river by a handsome three-arched aqueduct, now without its parapet. An embankment with buttresses and stone walls continues to the junction with the Bridgwater & Taunton; there is now a garden at the end of the canal with the towpath running through it. The house may have been the old toll house. There are no remains of the stop lock and Bridgwater & Taunton towpath bridge.

The Bridgwater & Taunton Canal is one of the very few surviving West Country canals. It too was bought by the railway but managed to continue trading until the beginning of this century. Its swing bridges were replaced by fixed bridges in the Second World War, with consequent limitation of headroom. Now

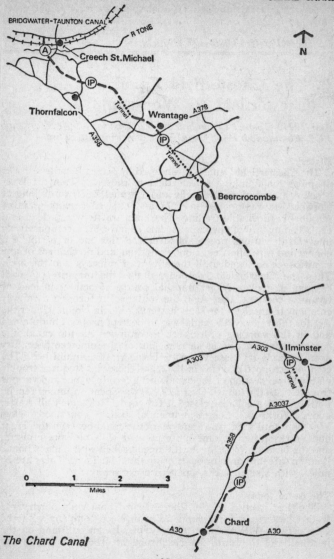

CHARD CANAL

BRIDGWATER-TAUNTON CANAL

R TONE

N

(A) Creech St.Michael

(IP)

Thornfalcon

Tunnel

Wrantage A378

(IP)

Tunnel

Beercrocombe

A303

A303

Ilminster

(IP)

Tunnel

A3037

A358

(IP)

A30

Chard

A30

0 1 2 3
Miles

The Chard Canal

13

owned by the BWB, the canal is open to navigation by small craft only.

OS sheet 193 (1:50,000) 177 (1 inch).

3. The Chesterfield Canal (Chesterfield to Worksop)

Act: 1771. Opened throughout: 1777. Disused above Worksop: 1908. Abandoned: 1962. Canal below Worksop still in use.

History

The Chesterfield Canal was surveyed by James Brindley. It was 46 miles long, linking Chesterfield to the river Trent at West Stockwith, with sixty-five locks and a tunnel 2,850 yards long at Norwood. It was opened in 1777 and enjoyed seventy years of reasonably profitable trading. The main traffic was in lead and coal. The history of the canal seems to have been comparatively uneventful; various proposals to extend the line or to make a connection with Sheffield came to nothing and no changes of any significance occurred until the advent of railways into the area. Then the Chesterfield company, unlike the majority of canal companies, took a rather unusual course of action; instead of trying to compete, they took the initiative in forming a railway company themselves, the Manchester & Lincoln Union. Under the Act for this railway, the canal was to be kept properly maintained and in full working order. Trade continued on the canal, the company itself acting as carriers, but as the nineteenth century drew towards its close the waterborne traffic diminished. The company stopped carrying in 1892, and there was frequent trouble on the upper section of the canal owing to subsidence and various troubles with the tunnel. Over £20,000 was spent on tunnel repairs until in 1908 a collapse closed it for good. This ended all traffic above Worksop. Some commercial traffic continued below Worksop until 1955. In recent years the canal between the Trent and Worksop has become quite popular with pleasure cruisers, and the basin at West Stockwith, once packed with narrow boats, is now a safe mooring for boats using the tidal Trent. There are 26 miles, with sixteen locks, open as a cruiseway.

The canal today

The 'lost' section of the Chesterfield Canal runs northwards from **Chesterfield** for several miles in close company with the Chesterfield — Rotherham railway line. It can be found on the north side of the town, due east from the crooked spire, by the

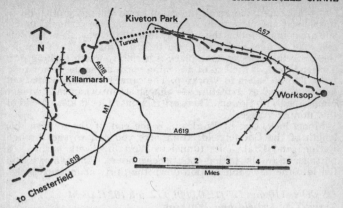

The Chesterfield Canal (Chesterfield to Worksop)

A619 to Staveley. A turning off this road, B6050 or Lockoford Lane, leads you to the canal if you do not want to seek it out in the Chesterfield traffic. You can follow the towpath southwards until the canal merges with the river Rother. The canal locked down from Chesterfield; the descent ends by the entrance to **Staveley Ironworks,** which were served by the canal in its prime. From here northwards remnants of the canal and fourteen narrow locks may be found in a tract of railway-dominated desolation on the western side of **Killamarsh.** Access hereabouts is difficult, however.

A mile north of Killamarsh, near the Angel Inn, A618 crosses the canal, which has now swung to run almost due east. The **Norwood flight** of thirteen locks is on the east side of the road, with a white house, formerly the Boatman Inn, at the bottom. This flight, consisting of three staircases of three locks each and one of four locks, is being cleared by the Chesterfield Canal Society as a site of great historical interest. At the top is the bricked-up and rather insignificant portal of the long **Norwood tunnel,** with M1 crossing the top not far away.

For the eastern portal of the tunnel, you must take B6059 through Kiveton to the bend by **Kiveton Park** station. You can find the canal south of the station and follow it westward to a deep cutting and eventually to the portal, also bricked in. This end of the tunnel was extended by about 250 yards when the railway was built nearby. The Canal Society is clearing the towpath from the portal to Dog Kennel bridge and proposes to run a trip boat on the summit pound when the BWB has completed a dredging programme.

15

From here to Worksop, the canal keeps on the south side of the railway and close to it. Again, access is quite difficult; unless you are going to follow the canal by foot all the way, there are only two road bridges which take you to it. In this length of 5 miles there are thirty locks. One place is certainly worth seeking out: **Turner Wood,** at the end of a turning north along the minor road from Thorpe Salvin to Worksop. The canal has been cleaned out here and used as a feature — almost as an ornamental village pond, lined with flowers. There are remains of locks at each end of this minute village.

Morse lock, east of Worksop, is the end of the unnavigable length of the Chesterfield; from here you can voyage through Retford and Drakeholes tunnel to West Stockwith basin — a passage, or a walk, taking about twelve hours. Some water is still fed into the navigable section down the thirty abandoned locks.

OS sheets 119 and 120 (1:50,000), 112 and 103 (1 inch)
Train: Chesterfield and Worksop

---

4. The Dorset & Somerset Canal

Act: 1796. Never opened.

History

The Dorset & Somerset Canal was envisaged as a 49-mile long waterway running south from a point on the Kennet & Avon Canal between Bradford-on-Avon and Widbrook, via Frome, Wincanton, Stalbridge and Sturminster Newton to Shillingstone Okeford, near Blandford Forum. There was also to be a branch from Nettlebridge, meeting the main line at Frome. The original intention was to provide a link between Bristol and Poole, but this was frustrated by the opposition of various landowners. Hence the choice of the southern terminus. In the event, however, this had no significance. The Act provided for the branch to be made first. About 8 miles of the branch had been constructed by 1803, when the money ran out. The raising of more funds was authorised but they were not forthcoming, and after work had stopped it was never restarted.

· It had been hoped that the branch would prosper from the carriage of coal and estimated that it would cost £30,000 to complete. What was built, which included two aqueducts, several bridges and part of a tunnel, cost in all £66,000 and exhausted the money that that had been subscribed. Much was spent on the Balance Lock, invented by James Fussell, a local ironmaster. This consisted of two chambers divided by a central wall. In each chamber there was a watertight box, into which a boat could be

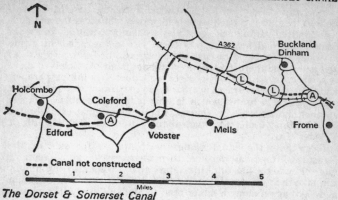

The Dorset & Somerset Canal

floated. By a system of wheels and chains, as one box ascended the other descended, the difference in height being about 20 feet. One of these locks was completed and demonstrated in action in 1800. The committee were so impressed that they decided to build five more, but none of these was finished.

The surveyor for the Dorset & Somerset was William Bennet, who was connected with several abortive canal schemes and who seems to have been consistent in under-estimating costs. When work on the D & S stopped he was employed for a time by the Somersetshire Coal Canal Company and then by the Kennet & Avon. Others connected with the D & S found other occupations, and the scheme remained dormant for over twenty years. There was a revival of interest in 1825 when it was proposed that the line, or some part of it, might be completed either as a canal or as a railway, but it came to nothing. In his recent history of this project, Kenneth R. Clew says that some parts of the canal still held water in 1840 and he reports a story that once barges had navigated between Edford and Coleford. But in the mid nineteenth century parts of the line were filled in and some of the land was sold off. As a sad postscript, the canal company's records were destroyed by bomb damage at Wincanton.

The canal today

Although modern maps do not show the remains of the D & S, exploration of them is not especially difficult. Where the branch was planned to begin at Nettlebridge, on A367, no work seems to have been done. The best place to begin is on the road running

2   17

south from Holcombe, by the Duke of Cumberland at **Edford.** Behind the pub is a packhorse bridge over what was the canal. A few hundred yards to the west was a basin at **Stratton Common.** Beside the road near the pub there is a depression with some masonry lining. Heading east, the course can be followed via stiles and gates, and there is a length of embankment. At **Coleford** there is a large, two-arched ivy-covered aqueduct, lacking its parapet, and there are other traces near **Goodeaves Farm,** where some work was done on a tunnel which is now indistinguishable. There is another short stretch at **Vobster,** where the original canal bridge can be found under the road about a third of a mile south of Vobster Cross.

From here the canal swings north to the far side of the Radstock-Frome railway line, then turns eastward and remains quite close to the line until the approach to Frome. If travelling by car, take the road running north from Vobster to A362, turn right and right again along a minor road to Mells. When you come to a Y junction, leave the car and walk parallel to the railway line towards Frome. The line of the canal is obvious; follow it along **Barrow Hill** and you will come to the site of Fussell's demonstration Balance Lock, now a hole in the ground with shrubs and trees growing about it. Continue in the same direction, but diverting if farmland demands it, as far as a small wood. Here the channel becomes clear and leads you to four sizable holes, with masonry remains in three of them. These were the unfinished balance locks; five of them were planned to drop the canal about twenty feet a time before levelling off for Frome.

It is easiest to return to your car and rejoin A362 through Buckland Dinham, stopping near the bridge over the railway. A stile on the south side of the bridge brings you, if you persist through the nettles, to a clearly defined stretch of canal bed and then to the solidly built **Murtry aqueduct,** taking the canal over the stream through Vallis Vale. This aqueduct is well worth examining. If you return to where A362 crosses the stream you can see what is left of the old road bridge below on the west.

It was intended that the branch should meet the main line on the north-west side of **Frome.** The only surviving item is a length of stone walling, to retain an embankment, on the north of Whatcombe Farm near the river Frome. If you cannot find it, ask for the Roman Wall.

OS sheet 183 (1:50,000), 166 (1 inch)
Train: Frome

5. The Grand Western Canal

Act: 1796. Opened: Tiverton — Lowdwells 1814; Lowdwells — Taunton 1838. Abandoned: Lowdwells — Taunton 1864; Tiverton — Lowdwells 1962. Length: 24½ miles.

History

A canal link between Bristol and Exmouth was one of the most obsessive dreams of canal promoters in the south-west of England. A waterway wide enough to take 15-foot beam sailing barges was contemplated, with branches to Tiverton and Cullompton. The promoters consulted a number of engineers, who varied each others' proposals and increased the estimates; meanwhile prices rose and, owing to the war with France, money was in short supply. Two undertakings were planned; a canal from Bristol to Taunton and another, the Grand Western, from Taunton to Topsham on the Exe south of Exeter. The first scheme hung fire for many years. An Act was obtained in 1811 but was amended in 1824, to result in the construction of a canal between Bridgwater and Taunton a few years later. This canal is now owned by BWB and open today for light craft only.

The Grand Western's Act was passed in 1796, allowing for a main line 36½ miles long with two branches. The estimate quoted was that of John Rennie, a total of £211,875. However, on account of the difficulties mentioned, nothing was done for several years until some of the Kennet & Avon shareholders took up Grand Western shares in the hope that this line, when completed, would also bring added traffic to their own, making possible an eventual inland water line between London and Exeter. In 1810 work began; not however at Taunton or Tiverton or anywhere near Exeter but in the fields near Holcombe, on the summit level of the canal and on the Tiverton branch. In this rural setting Sir George Yonge cut the first turf before an admiring audience of gentry and peasantry. Money and cider were distributed to the latter, while the former retired to enjoy the hospitality of Holcombe Court. The reason for beginning with the branch was that it would provide a direct route from Canonsleigh quarries to Tiverton and revenue from carrying lime and limestone would quickly build up. Rennie's estimate for constructing the branch was a little over £86,000. But it took four years to cut the 11 miles of canal — which had no locks—and it cost no less than £224,505. To add to the gloom of the situation, the tolls which had been anticipated at £10,000 *per annum* averaged only £600 until the 1830s, when they managed to crawl over the £1,000 mark.

19

In this situation the canal company took no further action over the main line for several years after the branch was opened. When the Bridgwater & Taunton Canal opened in 1827, however, proposals for continuing with the Grand Western were made. The engineer James Green reasoned that linking the branch with Taunton would make it profitable. With economy in mind, he suggested a tub-boat canal and produced an estimate of £61,324 for 13½ miles of narrow canal with seven lifts and an inclined plane, raising it a total of 262 feet. The canal company approved but placed a limit of £65,000 on expenditure, and cutting began in 1831.

Seven years later, Taunton and Tiverton were at last connected by water. Green's lifts, constructed on similar principles to Fussell's Balance Lock on the Dorset & Somerset, were troublesome to operate and modifications had to be made to the design. Furthermore, despite his experience with inclined planes on the Bude and Torrington canals, Green could not make the Grand Western incline at Wellisford operate at all. In 1836 the company sacked him and asked W. A. Provis, an experienced engineer, to advise. He accepted the lifts and discovered what was wrong with the incline. This was designed on the bucket-in-the-well system. Each track of the incline was linked to a bucket which, filled with water, descended in a well to provide the power by which a loaded boat ascended the incline. When the bucket reached the bottom (and the boat reached the top) a valve opened to release the water. Green's buckets were not big enough to take the weight of the 8-ton tub-boats. The committee raised enough money to buy a steam-engine to power the incline and repair one of the lifts, having borrowed some of it from their new superintendent, Captain Twisden, an elderly retired naval officer. The line was opened on 28th June 1838.

Coal and lime were the main cargoes on the Grand Western, carried in trains of four to eight tub-boats drawn by one horse. Tolls moved up towards £5,000 *per annum* but ominously much of the increase was due to the carriage of materials for the Bristol & Exeter Railway. When this opened in 1844 the tolls fell away, dropping below £1,000 *per annum* five years later. Competition could not be sustained; in 1853 the Grand Western company leased the canal to the railway for £2,000 *per annum*, with an option to purchase. This was taken up in 1864, when the canal was bought for £30,000. Three years later the Taunton-Lowdwells section was closed. Limestone traffic continued on the rest of the line, bringing in a few hundred pounds a year, until 1924. In 1962 this Tiverton-Lowdwells section was abandoned by BWB, but in 1971 it was transferred to Devon County Council with a grant of £38,750. The Council are improving the condition of the waterway and the towpath but have banned all forms of power boating.

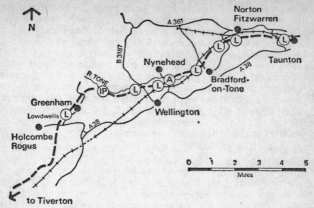

The Grand Western Canal

The canal today

Canal Road, **Taunton,** is the appropriate place for beginning an exploration of the remains of the Grand Western. From the bottom of the road you can see the site of the junction of the Grand Western with the Bridgwater & Taunton, where there used to be a stop-lock. Behind a row of cottages in the road is the railway goods yard; part of the retaining wall of one of Green's lifts has been incorporated into the boundary wall. The down relief line lies on the course of the canal, raised 23½ feet above the Bridgwater & Taunton by the lift. Although railway and canal soon part, the canal has been generally obscured for the next mile, but **Silk Mills bridge** survives on the west of Taunton. Out of town, the canal line runs between the river Tone and the Taunton-Exeter railway line for about 3½ miles. It can be found by crossing the footbridge at **Norton Fitzwarren** railway station and heading across the field. Walking westward along the depression you soon come to a change in level, the site of the **Norton lift** which raised the canal 12½ ft. You can continue walking to the site of the next lift, **Allerford,** a little over a mile onwards; or return to your car, turn on to A361 heading west and take the first left. Turn left at the Victory inn and you will pick up the canal on the far side of the level crossing, through some white gates. Allerford lift had a rise of 19 feet, and you may notice some masonry fragments.

The next point of interest is by **Trefusis Farm,** west of Bradford-on-Tone, where there is a road bridge. Continue on the road to East Nynehead, under the railway. In about half a mile a **track on the left leads to Lift Cottage,** which has considerable

21

remains of the 38-foot Trefusis lift in its garden. The canal is now on the north side of the railway line. Continue through **East Nynehead** to a T junction, turn left and then left again after about a mile. There is a cottage about 400 yards short of the railway bridge. This was a wharf cottage; if you walk a third of a mile eastward you come to a single-arch iron trough aqueduct which took the canal over the Tone. Back on the road, follow the same line through the trees to the west. Soon you come to the 24-foot **Nynehead lift,** which has been cleared out and is in the best state of preservation of all the lifts. To the right are the foundations of the lift cottage. A little further along, a fine single-arch aqueduct carried the canal over the drive to Nynehead Court.

The site of the 18-foot **Winsbeer lift** can be found close to the Tone by taking a path leading west off B3187, 1½ miles out of Wellington. There are ruins of the lift cottage but no masonry remains of the lift itself. The easiest way of finding the **Wellisford inclined plane** is to continue walking for two miles keeping close to the river. Otherwise you must make your way through winding country roads to Thorne St Margaret and ask for Bughole Lane, which takes you to Incline Farm. This is at the top of the 81-foot incline; if you ask, you may be shown the engine-house and various other traces.

The next place to make for is **Greenham,** about 2 miles south-west of Incline Farm. The canal bed is discernible at the south end of the village. Follow it south-westward, being wary of marshy ground and encroaching undergrowth. You will come to an accommodation bridge; the wilderness on the far side covers the remains of the **Greenham lift,** with the greatest rise of 42 feet. This lift collapsed during construction and had to be rebuilt. At the top is the lift cottage, now inhabited, so permission for exploration should be sought. From the top the canal continues for half a mile to a minor road, which once it crossed by an aqueduct. Stonework of the abutments can be seen underneath the ivy. You are now at **Lowdwells.** On the far side of the road the Devon County Council's section — the old Lowdwells-Tiverton branch — begins, by Lock's Cottage where there used to be a lock.

There is no problem about tracing the rest of the Grand Western. You can walk the length of the towpath to Tiverton or follow it by car along minor roads from Greenham to Westleigh and thence to Sampford Peverell. The minor road forking left by the church runs alongside the canal. This joins A373 near Halberton. Turn left after Halberton church as far as the railway; then turn right to see the aqueduct carrying the Grand Western over the Tiverton branch line. Turn right at the next T junction for Tiverton; the canal basin is on your right as you enter the town. On your way look out for the short tunnel under the Greenham—

Holcombe Rogus road junction, the wharf at Sampford Peverell and the embankment on the north side of Halberton. The canal from Lowdwells to Tiverton is clearly indicated on the OS map. Devon County Council publish descriptive leaflets of this section.

OS sheets 193 and 181 (1:50,000), 177 and 164 (1 inch).
Train: Taunton, Tiverton Junction. Tiverton by bus from Exeter.

6. The Grantham Canal

Act: 1793. Opened throughout: 1797. Abandoned: 1936. Length: 33 miles.

History

> The goddess of peace shall her blessings unfold,
> Then open your treasures and pay down your gold,
> Our trade it will flourish — then freemen be bold,
> To finish our new navigation,
> To finish the Grantham Canal.

So sang the promoters of the Grantham Canal, on hearing that they had obtained their Act of Parliament authorising them to raise money for their proposed waterway from Grantham to the Trent at West Bridgford, Nottingham. The gold was paid down; William Jessop was appointed to supervise the work, with James Green and William King as engineers for the western and eastern sections respectively. By the end of 1797, £120,000 having been spent, the canal was open. The cost per mile was comparatively low; there were eighteen broad locks going down to the Trent but no large-scale engineering works were needed.

For several decades the canal was moderately prosperous. Coal, coke, lime, various manufactured articles and groceries were distributed along the route to Grantham and agricultural produce was brought back. The only source of trouble was near the village of Cropwell Bishop where the canal ran across gypsum beds and frequently sprung leaks, flooding the workings.

The canal company maintained moderately high tolls, thus exposing themselves to railway competition. They decided not to fight it out and agreed to sell out to the compendiously named Ambergate, Nottingham, Boston and Eastern Junction Railway when their line from Nottingham to Grantham was opened. This occurred in 1850, but the railway company refused to pay up, partly because it was seeking shelter in the bosom of the Great Northern and partly because it could not afford to. It took four years and several lawsuits before the money appeared. The railway company tacked '& Canal' on to their name before shortening themselves and the extent of their proposed operations to the Nottingham & Grantham Railway & Canal Co. In 1861 the company merged with the Great Northern, eventually to become part of the LNER.

Trade fell away in the second half of the nineteenth century. Regular traffic ceased in 1917. Seasonal trade in grain continued for another few years, and a few pleasure cruisers appeared, but in 1936 the LNER obtained an Abandonment Act, arguing that the lock gates needed replacing and the navigation did not warrant the expense. The gates in fact survived into the 1950s, when many of them became prey to vandalism. British Waterways Board were obliged to keep the canal watered, but sold several of the bridges to Nottingham County Council, which levelled them in the interest of road safety. Many lock-keepers' cottages were vandalised and then demolished and the locks were converted to weirs.

In 1969 the Grantham Canal Society was formed and successfully opposed moves by BWB and the Trent River Authority which would have further worsened the state of the canal. Plans for restoration were drawn up and various minor improvements made. The Society, now called the Grantham Canal Restoration Society Ltd, aims at restoring the waterway to full cruising standards throughout and has formed a trust to organise the co-operation of the local authorities, BWB and the volunteer labour on the canal.

The canal today

Wandering amiably through the Vale of Belvoir the Grantham Canal is a pleasantly rural waterway for almost all of its length. It has lost its basin in **Grantham**, but can be found along Earl's Field Lane on the west side of the town. You can walk along the towpath as far as the A1. Much of the towpath has been cleared along the whole length and most of the bridges, until Nottingham is approached, remain. Access can be obtained from **Harlaxton**, where the canal runs through a narrow cutting, **Denton** and **Woolsthorpe** — follow the lane to the Rutland Arms (known locally as the Dirty Duck). Behind the pub is an original canal building — the old carpenters' shop. The villages of **Redmile, Barkestone, Plungar** and **Hornby** are on or quite close to the canal; so are **Hose, Hickling** and **Kinoulton** as you travel west. Hickling basin can be seen, with its late eighteenth-century canal warehouse in sadly dilapidated condition. North of Kinoulton is a bend called Devil's Elbow; the canal heads northwards from here, close to a minor road to Cropwell Bishop. Here is the leaky stretch referred to previously. The A46 (Fosse Way) crosses the canal at Berry Hill. Between here and Nottingham the waterway is in a sorry state and the Restoration Society has proposed making a new

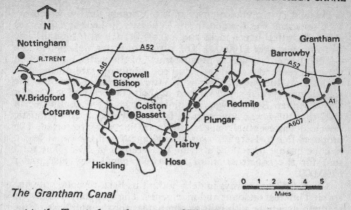

The Grantham Canal

cut to the Trent along the course of Thurlbeck Dyke.

The Grantham Canal, marked throughout on OS maps, is easy to find yet takes the discoverer through countryside that is little frequented and has a feeling of remoteness. It is, however, threatened by the proposals to mine in the Vale of Belvoir; if this happens, subsidence is likely to endanger the future of the canal. In connection with the mining plan, the BWB is considering a new cut from the Trent near Newark into the Vale and this too might pose problems for the continued existence of the old canal.

OS sheets 130 and 129 (1:50,000), 122, 113, 112 (1 inch).
Train: Grantham, Nottingham.

7. The Herefordshire & Gloucestershire Canal

Act: 1791. Opened: Gloucester–Ledbury 1798; Ledbury–Hereford 1845. Abandoned: 1881. Length: 34 miles.

History

The Herefordshire & Gloucestershire Canal was a locally inspired undertaking. Other proposals had been put forward for connecting Hereford with the Severn and for improving the Wye, but the opening of new collieries at Newent encouraged the promotion of a canal. Josiah Clowes surveyed the line and it was expected that it would cost about £100,000. Nothing much happened for two years after the Act was obtained; the line was then re-surveyed and, although money was slow to come in, work started in earnest. The long Oxenhall tunnel proved difficult to cut and cost over a quarter of the estimate for the whole. In 1798 the canal was opened from Gloucester to Ledbury, 16 miles in length with thirteen locks and a short branch at Oxenhall. The cost was now over £100,000. The supply of water was inadequate — the Frome river was to provide much of the water but this had not been reached — and the canal could only operate during part of the year. Income was low with little prospect of increase; the proprietors showed small interest in their concern and it was not until Stephen Ballard was appointed clerk to the committee in 1827 that there was any renewed impetus. He took an optimistic view and was eventually successful in getting work restarted. The proposed line to Hereford was altered again and the canal began to edge in that direction, arriving in Hereford in 1845. The total sum for the construction of the whole canal was just under £250,000.

The H & G was always in debt and, at its best, revenue only just covered interest on loans and mortgages. The company was always ready to negotiate with railway companies for the purchase of the line. After much delay, the GWR took it over for £5,000 *per annum*. In 1881 the Gloucester-Ledbury section was closed and drained, and railway track was laid over the straighter parts of its course. The Ledbury-Hereford section simply fell into disuse.

The canal today

The site of the basin in **Hereford** is opposite the front of the railway station at Barr's Court. Little can be traced in Hereford itself; there is a 440-yard tunnel under **Aylestone Hill** but the portals are on industrial property and may not be accessible. About two miles north of the basin the canal swings north-east. An aqueduct took it over the river Lugg near **Shelwick Green**; the remains of the piers may be visible when the water level is low. Near Hereford the bridges have not survived. You can make out the site of the wharf at **Withington Marsh**, where the canal was crossed by A465; there is a group of buildings comprising the wharfinger's house ('William Bird, Wharfinger' used to be visible on the end wall), a weighing house and a canal cottage. On the road half a mile north of Withington church a bridge carries the date 1843, and a garden on the other side incorporates a lock. If you return to Withington and take the next left you will come to another wharf house at **Kymin,** where the course of the canal is discernible. Drive on to A417 and turn right on to A4103 at Newtown. You cross the canal in less than a mile. Continue for a few hundred yards and take the left turn to Monkhide. This road soon crosses the canal twice, both bridges surviving. The first is a sharp skew bridge; by the second there is a wide stretch of water, with a cottage on the bank, tree-lined and beautiful. It lasts for a little over 200 yards; then the water ends and there is only a dry ditch through the trees.

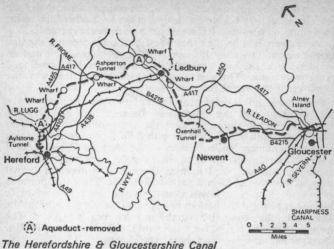

(A) Aqueduct - removed

The Herefordshire & Gloucestershire Canal

Make your way to A417, about a mile south of **Stretton Grandison.** There is a canal warehouse (the word 'salt' may still be just legible) on the west side of the road; opposite the line can be followed across a field to a bridge by the gates of Canon Frome school. It continues south-eastward to what is left of the north portal of **Ashperton tunnel** (sometimes referred to as Walsopthorne) about a quarter of a mile long. The turning to the east at Ashperton on A417 leads to a village cricket ground which is on the top of the tunnel. This road turns to run parallel to it; the south portal, in better condition, soon appears in a deep cutting. Originally it was intended that the tunnel should include the length of cutting as well, but it was shortened under Ballard's scheme for completing the canal. The road turns to cross the canal at the end of the cutting; soon the canal also turns north-east towards Swinmore Common, then east again for Staplow on A4154. Here a canal cottage stands near the road. For the next half-mile the canal runs behind Prior's Court and is accessible from the minor road turning left half a mile south of **Staplow.** If you can fight your way along this stretch you will find two small aqueducts, one over the Leadon, the other over a stream, and the remains — or fragments — of a lock gate (or stop gate). Return to the road; on the opposite side is the site of a wharf.

The canal now runs south for **Ledbury,** with an intact bridge on the road between A4154 and Wellington Heath. The main road

27

crosses it as you approach the town. The course passes under the two-arched railway bridge and keeps on the west side of the town; the swimming pool is on its line. From here on railway tracks, long removed, were laid on the canal. West of the town centre the line curved west and then south again, crossing A449 by what is now an engineering works but was originally a long, well-proportioned canal warehouse. There is also a wharf house; this was the terminus of the canal for many years.

From here until a point near Tillers Green canal remains have been obliterated by the old railway. Between here and Dymock they diverge, the railway taking the more direct route. From B4215, just south of the junction with B4024, you can see remains of an embankment on both sides of the road. In **Dymock** canal and railway coincide again, but half a mile south they part. The railway looped westward to pass through a cutting; the canal plunged into the 2,192-yard **Oxenhall tunnel.** The north portal is in private grounds at Boyce Court, on the north side of M50. Silting has brought the bottom nearly up to the top of the arch; the stone facing has gone and crumbling brickwork is exposed. The horse path can be traced for a few yards, with a stone footbridge. From the motorway you can see the spoil heaps.

For the south portal take B4215 over the motorway and turn second right. In half a mile this road crosses the canal — a wide, shallow ditch full of trees and scrub — and you can follow it northwards to the tunnel, this portal being in far better condition, retaining its stone facing. Return southwards, back across the road and you can follow it to the next minor road at **Oxenhall.** On the east side is the Furnace Pond, on the west concealed in the marshy foliage the junction with the Oxenhall branch. As the branch is rather higher than the main line there was probably a lock here, but no traces have so far been found. The branch can be clearly seen curling round to the south of Oxenhall church. It has recently been established that it continued eastward for about a mile across the fields and a brook to collieries north of Kilcot.

The main line carries on, with the remains of two locks and an inhabited lock cottage, across an aqueduct over the Ell Brook and is lost in a timber yard on the north side of **Newent,** where the railway rejoined it near the ruins of the station. The top lock by the cottage has its chamber in fairly good condition; these locks could take boats 70 feet long and up to 7 foot 6 inch beam.

Between Newent and Gloucester there is not much to be seen; apart from some of the sharper bends, the railway has eradicated most traces of the canal. The H & G locked into the Severn in the grounds of the present hospital on the east side of Over. For many years it was possible to discern its course across part of Alney Island, between two branches of the Severn, even though this was

used for only a few years. But this has disappeared beneath new road-building schemes and other developments.
OS sheets 149 and 162 (1:50,000), 142 and 143 (1 inch).
Train: Hereford, Gloucester.

8. The Huddersfield Narrow Canal

Act: 1794. Opened: 1811. Abandoned: 1944. Length: 19⅞ miles.

History

Many eccentric and ambitious projects flowered in the 1790s, a decade in which, to canal promoters, nothing seemed impossible. The majority of these originated in the south and west of the country where promoters and their engineers seemed to display more imagination than practicality. One is rather surprised to find such schemes in the harder-headed north, and it is true that there were comparatively few of them. One such, however, was the Huddersfield Narrow Canal, one of three completed projects to take waterway traffic across the Pennines and provide a link between the east and west coasts.

The Huddersfield Narrow joined Manchester, via a short length of the Ashton Canal, to Goole and Hull via Sir John Ramsden's broad canal, the Calder & Hebble and the Aire & Calder. It took the most direct route across the hills, being 13 miles shorter than the Rochdale Canal authorised in the same year but completed seven years sooner, and well over a hundred miles shorter than the Leeds & Liverpool Canal further north. This direct route, however, involved no fewer than seventy-four locks and the longest tunnel on the waterway system, as well as two other tunnels and eventually ten reservoirs. It reached a height of 656 feet above sea level, and cost about £400,000 to construct. Benjamin Outram was the first chief engineer.

Some of the canal was opened in 1799 but it was another twelve years before the tunnel was completed at Standedge and frequent financial injections were required. Standedge tunnel cost about £160,000; it was about 9 feet wide and 9 feet high and 5,456 yards in length, being later extended by 242 yards when the railway was built. The canal's locks could only take narrow boats, which caused difficulties for through traffic off the wider navigations to the north-east. There were problems with water supply and new reservoirs had to be made; maintenance was expensive and there was criticism of bad workmanship of the locks. Many traders preferred to use the Rochdale Canal and receipts on the Huddersfield were never high; in several years no dividend could be paid. In the 1840s railway competition made itself felt; reductions of tolls attracted some more trade but income declined.

In 1845 the canal was sold to the Huddersfield & Manchester Railway for a little over £180,000. This railway became part of the London & North Western two years later. In 1849 a railway tunnel was constructed at Standedge. Trade did not desert the canal, however, although the tunnel suffered from rock falls and little traffic passed over the summit. In the early 1890s a double-track railway tunnel was cut; it was then that the canal tunnel was lengthened at the Diggle end. Tonnage continued to be carried until the Second World War; then apart from half a mile at Huddersfield the canal was abandoned by an LMSR Act in 1944. The Huddersfield Narrow still provides water for industry and the tunnel serves to drain the railway tunnel above it, but the locks are unusable.

The canal today

The southern terminus of the Huddersfield Narrow is at **Dukinfield** where it joins the Ashton Canal, which is part of the 'Cheshire Ring' recently reopened to navigation. A warehouse belonging to the Huddersfield company was built at the Ashton's basin at Piccadilly, Manchester, about 6 miles west. The Huddersfield is shown on the OS maps and its towpath can be walked. It disappears into a culvert in **Stalybridge,** where over 1,000 yards of canal and four locks have been obliterated. Emerging, it runs close to A635, though part of the line is obscured by Hartshead power station. The road crosses the canal on the north of **Mossley;** then south of Mossley and east of the road is the 200-yard **Scout tunnel,** its portals now bricked up.

Continuing by road, take B6175, which crosses canal, river Tame and railway, to A670 through **Saddleworth.** You are now approaching the summit and the canal has ascended through thirty-one locks by the time you arrive at **Diggle,** half a mile east of A670. You drop down hill to the railway line and look for a phone-box. The portal of **Standedge tunnel** is easy to find; note the date 1893, the year in which it was extended.

Air-shafts over Standedge mark the line of the canal and railway tunnels below. The northern portals of all these are at Marsden, on the north side of A62. Make for **Marsden** station and the Junction Inn. Near here is a maintenance yard and a bridge over the canal from which you can see the locked and barred entrance to the canal tunnel, the two single-track railway tunnels and the double-track one which is still used. Legging was the rule on the canal tunnel, and the passage took about 3½ hours. Traffic was one way, boats only being allowed to enter at stipulated times. For a few years a steam tug operated, but it was withdrawn in favour of leggers who seem to have operated until the end of horse-drawn boats. Vivid descriptions have been published of voyages through the tunnel in recent years, but it is now

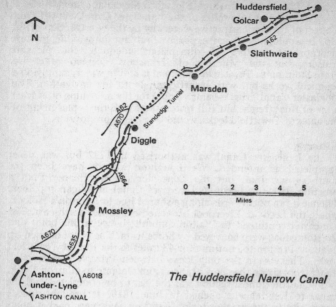

The Huddersfield Narrow Canal

considered too dangerous for visitors to be allowed through.

From Marsden the canal descends through forty-two locks. It runs parallel and close to A62 on its northern side. It keeps close also to the river Colne and you can find several single-arch aqueducts as it crosses and recrosses the river. At **Slaithwaite** it has been culverted for a short distance, but it is easy to follow into **Huddersfield** and makes a fascinating walk with splendid views. A long-term and admittedly costly plan for the restoration of the canal to navigable standard has been drawn up by the Huddersfield Canal Society.

OS sheets 109 and 110 (1:50,000), 102 and 101 (1 inch).
Train: Manchester, Huddersfield.

9. The Lancaster Canal (Kendal to Tewitfield)

Act: 1792. Opened: (Kendal-Wigan) 1819. Abandoned: 1955. Open from Preston to Tewitfield bottom lock.

One of the most unfortunate losses to the canal system in recent years was the top 14½ miles of the Lancaster Canal, cut off from the rest of this very attractive waterway by the construction of the M6. It would have been expensive but not impossible to have incorporated bridges of dimensions adequate for pleasure cruisers, but the Ministry of Transport decreed otherwise. From Stainton to Tewitfield the canal is still in water, supplying a chemical works but with insufficient depth for navigation. The Lancaster Canal Trust is campaigning for the reopening of at least some of this stretch, but full restoration, with the replacement of the gates of Tewitfield locks, would be a costly proposition.

History

The Lancaster Canal was authorised in 1792 but was never completed as intended. Westhoughton was to have been the southern terminal, but the Lancaster got no further than a junction with the Leeds & Liverpool Canal at Wigan top lock. Thence it ran northwards along a shared line to Johnson's Hillock, where the Leeds & Liverpool branched off to the north-east. The Lancaster continued to Walton Summit; between this point and Preston goods were conveyed by horse tram. The canal opened up again at Preston to continue for 43 miles to the foot of Tewitfield locks. These were the only locks between Preston and Kendal, apart from six on the short branch from Galgate to Glasson Dock.

The northernmost section was the last to be completed and the canal to Kendal was opened in June 1819. It brought speedy prosperity to the market town. Within a few years packet boats were running daily between Kendal, Lancaster and Preston, the fare being six shillings first class and four shillings second for the whole journey with a parcel service for one shilling an item. Fares were reduced sharply in the face of railway competition, but the boats provided a comfortable and efficient service until the canal company took over the running of the Lancaster & Preston Junction Railway in 1842 and withdrew the packet boats between Lancaster and Preston.

As a whole, the canal was never a strongly profitable concern, and dividends, when paid at all, were low. In 1885, after many years of negotiation, the proprietors sold the Preston-Kendal length to the London & North Western Railway. The canal passed to the LMSR in 1935. Until 1944 a few thousand tons of coal a year continued to be delivered to Kendal Gas Works, effectively now the terminus as the canal between here and the basin in Kendal was closed because of leakage. The 1955 Transport Act authorised the closure of the whole navigation. The top two miles nearest to Kendal were filled in and the next few miles drained. North of Tewitfield, the construction of M6 did the rest. The 42 miles of canal between Tewitfield and Preston, with Rennie's famous aqueduct over the Lune, escaped the rigour of the law and today provide popular cruising water with a number of boat-hiring firms. The branch to Glasson Dock is also open, except in periods of severe water shortage when the locks may be closed.

The canal today

It would be pleasant to suggest that the best way of exploring the northernmost section of the Lancaster would be by taking a boat to **Tewitfield** and then walking the rest. The towpath is a public footpath through most of the length. However, the first few miles are complicated by the motorway and its approach roads. The motorway runs parallel to the eight wide locks, a miserable sight where the few anglers seem especially disconsolate. A6070 keeps close to the canal and the turnings to the west lead to the watered section. There is a large number of bridges, particularly near **Holme,** and two aqueducts over minor roads. At Crooklands the motorway swings north-eastward to bypass Kendal; from here there is another mile and a half of water as far as a cross-roads south-west of **Stainton.** It is worth walking eastwards from here to a small aqueduct carrying the canal over St Sunday's Beck. The Lancaster aqueducts are solid stone constructions, this one and another on the dry section by **Sedgwick** built on the skew and of excellent workmanship.

Half a mile west along the dry section, the canal enters the 378-yard long **Hincaster tunnel,** taking it under the railway and a hill, emerging near a minor road on the north-west of the little village of Hincaster. The tunnel is built of stone and brick; the portals are stone-faced and the interior is stone below the waterline and brick above. The horse path can be followed under the railway and beneath two footbridges, coming down the hill to end behind a small group of cottages near the west portal.

From Hincaster the line of the canal is almost due north, parallel to the river Kent on its western side. It is possible to walk the course through the fields for about four miles until it peters out on the south side of Kendal. If travelling by car, you can pick up the line on the west side of Sedgwick and on the minor road half a mile west of **Natland,** where there is a good bridge. Here and there you find accommodation bridges isolated in fields, and there is another good stone bridge on a turning west off the Natland road as it approaches Kendal. Three engineers worked on the Lancaster — Rennie, Fletcher and Crosley — and the many bridges and aqueducts which remain are a tribute to their design and to the skill of the masons who built them.

OS sheet 97 (1:50,000), 89 (1 inch).
Train: Kendal

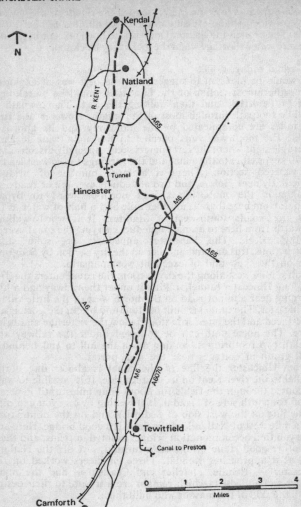

The Lancaster Canal (Kendal to Tewitfield)

10. The Leominster Canal

Act: 1791. Opened: 1796. Abandoned: 1858. Length: 18½ miles.

History

The original proposition for the Leominster Canal was exceptionally ambitious. As envisaged by Thomas Dadford junior who surveyed the route, it was to begin at Kington, run 12 miles east to Leominster, swing north to Woofferton, then head east again to connect with the Severn at Stourport, a total of 46 miles. A glance at the map will show the hilly nature of the countryside; indeed there were to be three summit levels, with lockage involving falling 496 feet and rising 48. Included were two long tunnels and three aqueducts. The idea was to open up Herefordshire to the manufactured products of the Birmingham area and to provide transport for agricultural produce in return. There were also collieries near Newnham which would be linked to Leominster by the canal. It may have looked very attractive on paper; in practice, however, it did not work. Like many other canal projects of the late eighteenth century, the Leominster Canal demanded extensive engineering works for which the money was not available. It was estimated that the line would cost at least £150,000, but less than half of this was raised. It is clear that had the works been completed they would have cost very much more than this anyway.

Work began soon after the Act was obtained in 1791, and in 1794 some few miles were open. Construction of the tunnels was giving much trouble; the 330-yard tunnel at Putnal Field caused great difficulty and the north end of the long Southnet tunnel fell in. John Rennie was consulted and was unimpressed by what he saw. He revised the estimates; the company obtained another Act in 1796 to raise more money, but it was not forthcoming. Some coal was being carried on the completed section from Marlbrook to Leominster, but not enough to alter the financial outlook. In 1803 work on cutting the canal stopped; there were various proposals to complete the line by tramroads and to connect the Leominster to other existing and projected waterways, but very little was done. For four decades some desultory trading carried on; then with the advent of railways into the area the proprietors began to negotiate to sell the canal. In 1846 the Shrewsbury & Hereford Railway offered £12,000 for the canal; it took eleven years and a lawsuit before the deal was completed. In 1858 the length from Leominster to Woofferton was closed and the rest followed in the next year. In 1860 the Tenbury Railway bought a section and used it for laying its track.

The canal today

Some earthworks intended for the canal may be seen in a field north of Kingland, near the river Lugg, but to explore this canal it is best to start at **Leominster** and head north. A mile out of Leominster on the A49 Ludlow road is the Coal Wharf house on the left, with the course of the canal behind it. It runs northwards, skirting the grounds of Berrington Hall on the west side, and is crossed by the railway line about 1½ miles north of Eye. If you continue along A49 and turn left for Orleton you come to a bridge over the railway. In another 50 yards you are on top of **Putnal Field tunnel.** By walking a short distance in each direction you can examine both portals and there is an attractive stretch of tree-lined canal on the southern side. The canal keeps close to the railway line as far as B4362. Find the Salway Arms, by the junction of this road with A49. Nearby is the derelict **Woofferton** station; across a field opposite the station you can find the canal again. Follow it to the right under the railway line and you will come to the north wall of one of three locks, where the heel post has rooted and grown into a tree. There are traces of the other locks, and a cottage that used to belong to the lock-keeper.

At Woofferton A456 branches eastwards off A49. On the left of this road is the track of the Tenbury Railway, coinciding with the canal. Soon the lines diverge, the canal curving north-eastwards on an embankment. This takes you to the remains of the aqueduct over the Teme, which survived until the last war when the central of its three arches was blown up, believe it or not, as an anti-invasion measure. Return to A456 and head for Kidderminster. The left turns off this road all cross the course of the canal, though not much is visible. About 6 miles further on there is an interesting stretch at **Newnham Bridge.** Turn left on to the minor road past the old railway station. A farm lane indicates the line of the canal. Follow it eastward; you will come to the site of a short tunnel filled in a few years ago. On the far side the canal crosses a field past two cottages; you can follow it until you come to a crumbling brick aqueduct over the Rea. Cross the aqueduct (with care) and continue for 2 miles as far as Wharf House at **Marlbrook,** where the canal ended. On this walk you may see the traces of several locks, and a restored lock-keeper's cottage. Wharf House itself is a handsome red-brick building. On either side of the central bay were docks into which boats could be drawn for repair. Coal was brought down to this point by tramroad and fragments of rail are frequently turned up in the grounds. The house is in private occupation. If the walk is too long, or too wet, Wharf House can be reached from a lane off A456, 2 miles from its junction with A443 to Worcester.

Further work on the canal was done to the east of Wharf House. As has been said, the north portal of the 1,250-yard **Southnet** (or

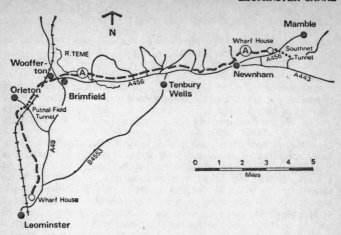

The Leominster Canal

Sousant) tunnel collapsed; you can pinpoint the site on the opposite side of A456 from the lane to Wharf House. The southern portal could be found in a field belonging to a farm up a left turn by the Nag's Head, three miles from Newnham Bridge towards Worcester on A443. It was of odd design, tall and narrow, and can never have inspired confidence. There is evidence that some further cutting was done towards Stourport, and that a start was made on the proposed 3,850-yard Pensax tunnel. But of the seventeen locks descending to the Severn nothing was begun and only one ceremonial spadeful was dug to connect the Kington, Leominster & Stourport Canal — to give it its full title — to the rest of the kingdom.

OS sheets 149 and 138 (1:50,000), 129 (1 inch).
Train: Leominster

37

11. The Rochdale Canal

Act: 1794. Opened: 1804. Abandoned: 1952 (apart from a short section in Manchester). Length: 33 miles.

History

Engineered by William Jessop, the Rochdale Canal was one of the most remarkable achievements in canal construction. With ninety-two broad locks in its 33 miles, it cost about £600,000 to cut. While the Huddersfield Narrow pierced the Pennines with its long tunnel at Standedge, thus providing itself with a built-in and troublesome bottleneck, the Rochdale strode over the hills, rising to about 600 feet above sea level at the summit. It linked the Bridgewater Canal in Manchester to the Calder & Hebble at Sowerby Bridge. The locks were built to approximately the same dimensions as those on the Bridgewater, 14 feet 2 inches wide, avoiding transhipment between those two navigations. The Calder & Hebble locks, however, were too short to take narrow-boats or barges, so cargoes bound for the north-east had to be transferred at Sowerby Bridge. But the wide Calder & Hebble (and Aire & Calder) boats could use the Rochdale locks for cargoes heading for Manchester and Liverpool; this gave the Rochdale a further advantage over the Huddersfield Narrow.

For many years the Rochdale traded with much success. By 1812 it was carrying well over 200,000 tons a year and in another ten years this total was doubled. The half million mark was reached in 1829 and a decade later the figure was approaching 900,000. Coal, corn and general merchandise were the principal cargoes. The canal company invested much of its profits in warehouses and reservoirs and was also able to provide a comfortable return to its investors.

The Manchester & Leeds Railway arrived in the Rochdale's area in the early 1840s and rate-cutting battles began. Agreement in the interests of both parties was reached in 1843 and was followed by several years of manoeuvring between the various railway and canal companies operating in the region. The Rochdale was prepared to sell itself to the Manchester & Leeds in 1847, but the arrangements failed. Eight years later four railway companies combined to lease the canal for twenty-one years at a rent of £37,652 *per annum*, enough to pay reasonable dividends and maintain the canal as a going concern. Less and less traffic was going over the summit, however, and much of the tonnage carried on the Rochdale used the short section in Manchester between the Bridgewater and Ashton canals.

Towards the end of the century, after the railways had lost their interest in carriage by water, the canal company began its own carrying department. They owned a total of sixty-eight boats of

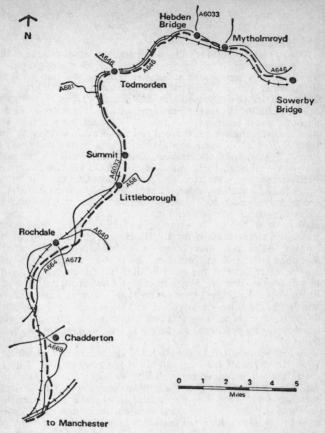

The Rochdale Canal

different types in 1892. The company gave this up in 1921 at a time when, although canal trade was continuing, it seemed to them that survival could only be attained through government assistance. Several of their reservoirs were sold to local authorities in 1923, but revenue from tolls was diminishing almost yearly. Apart from the connecting length between the Bridgewater and Ashton, the navigation was abandoned in 1952, although the

39

company remains in being, deriving income from sale of water, investments and the development of property.

The canal today

The Rochdale Canal is still a waterway, despite the damage that has occurred to it since the closure of the navigation. In Manchester it terminates at **Castlefield,** the junction with the Bridgewater Canal. The first mile has been restored and now forms part of the reopened Cheshire Ring, a circle of navigable canals including the recently restored Ashton. There are locks and a tunnel under Deansgate on this section, and the junction with the short Manchester and Salford Junction Canal, which was open from 1839 to 1875. The Rochdale and Ashton canals met north of Piccadilly station. The Rochdale's Dale Street basin is a car park.

A few years ago three miles of the Rochdale Canal through **Failsworth** were converted by Manchester Corporation into a water park, a shallow water channel with landscaped borders, at considerable expense. This has suffered disastrously from the depredations of vandals and provides a sad comment on the civilisation of the 1970s. From Failsworth the canal swings north through **Chadderton** and towards Rochdale, climbing towards the summit. A664 from Manchester to Rochdale crosses the canal at **Slattocks** and again at **Castleton.** South of Rochdale the canal is on the north side of the road behind the Arrow Mill. On this length are two fine bridges, Gorrell's and March Barn, reputed to be the first skew bridges built over a canal. There is a short branch into **Rochdale,** but unless you are walking the whole length of the towpath, which is possible, it is best to continue to **Littleborough,** a town owing its prosperity to the canal, and look at the Rochdale there.

In Littleborough, take the A6033 Todmorden road which leads you to the **Summit Inn,** with an old lodging house behind it. The canal has now ascended through fifty-six locks and the scenery is formidable. Canal and road keep close company into **Todmorden;** turn then on to A646 for Hebden Bridge, with the canal remaining on the south side of this road. In **Hebden Bridge** canal crosses river by a workmanlike four-arched aqueduct, found near the post office. Continue on the same road to **Sowerby Bridge,** where you can find the junction with the Calder & Hebble opposite the church.

Being close to main roads for nearly all of its length, the Rochdale Canal is easy to discover. Despite the destruction of some of the locks and the demolition of a few bridges — and the siting of a pylon in the middle of the channel at Chadderton — most of the canal is in good condition and provides good angling as well as a water supply for industry. Complete restoration would be difficult and expensive; so far, the Rochdale Canal Society has restored one lock on the West Yorkshire-Greater Manchester border and through job creation schemes the canal has been tidied up in Sowerby Bridge and improvements have been made in Rochdale itself, including the restoration of two locks. But the canal is now threatened by the M66 extension — the proposed Manchester outer ring road. If this is laid across the canal, the return of through navigation will be a virtual impossibility.

OS sheets 109, 103, 104 (1:50,000), 101, 95, 102 (1 inch).
Train: Manchester, Rochdale, Hebden Bridge, Sowerby Bridge.

12. The Shropshire tub-boat canals and the Shrewsbury Canal

Donnington Wood Canal. Act: none. Opened: c 1768. Disused by 1904. Length: 7½ miles (including 2-mile branch from Hugh's Bridge to Colliers End). Ketley Canal. Act: none. Opened: 1788. Disused by 1816. Length: 1½ miles. Shropshire Canal. Act: 1788. Opened: 1792. Abandoned in sections: 1857, 1913, 1944. Length: 10½ miles (including Coalbrookdale branch 2¾ miles). Shrewsbury Canal. Act: 1793. Opened: 1796. Abandoned: 1931 and 1944. Length: 17 miles.

History

The tub-boat canals of east Shropshire formed a connecting network of waterways serving the ironworks and coalfields of this heavily industrialised area. Although there was no physical connection with the Severn, there were wharves at Coalport where goods could be transferred to the Severn trows. Later in their history, a link was made with the Birmingham & Liverpool Junction Canal via the Newport arm which met the Shrewsbury Canal at Wappenshall. They were designed for tub-boats, rectangular floating boxes less than 20 feet in length, which linked to each other and were drawn by a horse. The Newport arm, however, took the conventional narrow-boats, while the Shrewsbury Canal's locks were long enough for narrow-boats but only 6 feet 4 inches wide.

The earliest of these canals, the Donnington Wood, was built by Earl Gower, brother-in-law of the Duke of Bridgewater and a business partner of Bridgewater's agent, John Gilbert. In time, Earl Gower became further ennobled as Marquess of Stafford and Duke of Sutherland, and his canal was also known by these two titles. It was intended to convey coal from mines on his Donnington Wood estate to a wharf two miles from Newport on the road to Wolverhampton. On this line there were no changes of level, but a few years later a branch was added connecting limestone quarries at Lilleshall with the main line at Hugh's

Bridge. This branch had seven locks. At Hugh's Bridge it was some 42 feet lower than the main line and at first goods were transferred by cranes through two shafts, being lowered to or raised from a short tunnel at the end of the branch. In the 1790s an inclined plane was constructed to effect the transfer. The branch continued to be used until the 1870s; a few years later only the section at the Shrewsbury Canal end was in use, and by 1904 trade on the Donnington Wood had ceased altogether.

About twenty years after the opening of the Donnington Wood, William Reynolds, the ironmaster of Ketley, was responsible for constructing three tub-boat canals. The Wombridge Canal, less than two miles long, joined Earl Gower's canal at Donnington Wood and, when the Shrewsbury Canal was opened a few years later, connected with that at the top of the Trench inclined plane. The Shrewsbury then took over most of the line of the Wombridge, which continued in use until about 1904.

Reynolds's second canal, the Ketley, was another short waterway, planned to supply the Ketley works with coal and ironstone from Oakengates. This canal is historically important as it was the first to incorporate an inclined plane to overcome the problem of change of gradient. It was double-track, the descending loaded boats, fitted into cradles, providing the power to raise the empty boats coming up. The Ketley Canal, which also had a short tunnel, served as a branch of the Shropshire Canal when the latter was opened in 1792, but became disused when the ironworks closed in 1816.

The Shropshire Canal, by far the largest of these undertakings, was sponsored by a group of local industrialists, including the Darbys and Lord Gower — now Marquess of Stafford — as well as William Reynolds, his brother Richard, and Thomas Gilbert, brother of Bridgewater's agent John. The £50,000 estimated cost was raised immediately and the line, which included three inclined planes, was completed over difficult country in 1792, four years after the passing of the Act. It joined the Donnington Wood, and a couple of years later the Shrewsbury also, at the foot of the Wrockwardine Wood incline. It headed south through mining country and the junction with the Ketley Canal to Southall Bank. The main line continued via the Windmill Farm and Hay inclines to wharves alongside the Severn at Coalport, while a branch swung off westward to Brierly Hill, above Coalbrookdale. At this point, vertical shafts, similar to but larger than those at Hugh's Bridge, were built for the transfer of goods by crane. Within a few years this method was replaced by a railway inclined plane.

The Shropshire Canal traded successfully until 1845. It then became part of the Shropshire Union amalgamation. In the next few years its condition worsened due to mining subsidence, and trade diminished. The London & North Western Railway bought

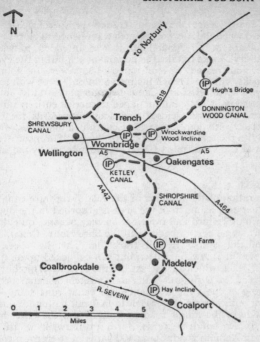

The Shropshire tub-boat canals and the Shrewsbury Canal

the canal in 1857 and closed it as far south as the Windmill incline. The Hay incline was last used about 1894, and only a short section serving the Blist's Hill furnaces survived into the twentieth century. It was not until 1944 that this was legally abandoned.

Many of the promoters of the Shropshire Canal were also involved in the Shrewsbury, designed to take coals to that town from the mines of east Shropshire. This was also intended for tub-boats, with eleven locks each long enough to take four tub-boats, with guillotine or lifting gates at the lower ends, and one inclined plane at Trench. It was cut from Shrewsbury, where it did not connect with the Severn, generally eastward through Berwick tunnel, over the Tern by an iron aqueduct at Longdon, past Wappenshall, where the Newport branch later connected with it, to Trench, where at the top of the incline it joined the Donnington Wood Canal by means of the Wombridge length which it bought. Josiah Clowes was the first engineer, being succeeded by Telford

43

in 1795.

For many years the canal traded prosperously. Dividends increased when the Newport branch was opened to Norbury on the Birmingham & Liverpool line. It was intended to widen the Shrewsbury's locks to take the narrow-boats, but in the event only two were so treated. Special narrow-boats were built to fit the other locks, but the Trench incline could only cope with tub-boats anyway. The Shrewsbury Canal became part of the Shropshire Union system in 1846 and managed to survive entirely until 1921. Then the incline was closed and soon afterwards the basin at Shrewsbury, although some traffic continued to the gasworks until 1931. With the remaining portions of the east Shropshire network, the rest of the Shrewsbury Canal was abandoned by the LMSR in 1944.

The canals today

Exploration of the remains of the east Shropshire canals is not altogether easy as much of the area is absorbed in the developing new town of Telford and information may become quickly out-of-date. It may be best to begin with the Shrewsbury Canal, of which most of the course is still traceable. This has been mostly obliterated in Shrewsbury itself but can be picked up on the south side of B5062, which turns east off the A49 to Whitchurch. The minor road to **Uffington** leads to a stretch of canal bed at the beginning of the village. Thence the canal runs south towards Preston; both portals of **Berwick tunnel** can be found, one on the south side of the minor road to Preston and the other behind a house a few hundred yards north of Berwick wharf on the Uffington-Atcham road. Some 400 yards of overgrown canal and towpath, with an accommodation bridge, lead to the latter portal, which bears the date 1797 on the keystone. Both portals are now bricked up. They are stone-faced, although the tunnel itself, which originally had a towpath built through it, was lined with brick. A ventilation shaft used to be visible part way along the 970-yard length, but this has now been demolished.

From Berwick wharf the canal turned north-eastward through Withington and Rodington, where it crossed the Roden by a three-arched aqueduct. It is possible that a lifting bridge may survive beside the road to Longdon. Continue towards Longdon and turn right on B5063; there is a Shropshire Union Canal Company warehouse at the site of **Longdon wharf** and a few hundred yards further on you can see on the north side the iron-trough aqueduct over the Tern. This was the precursor of the great aqueduct by Telford at Pontcysyllte on the Llangollen Canal. At first the canal was carried over the river by a stone aqueduct; this was demolished by floods and the iron trough was substituted under Telford's direction but using the original masonry abutments. It is

62 yards long and includes a towpath. It is possible that it will not remain on site much longer, but may be removed to the open air museum at Ironbridge Gorge.

The canal is crossed by A442 at Long Lane and then heads south-eastward across Eyton Moor. There is a lock about half a mile from Long Lane, on the land of a local farmer, and another on the north side of **Eyton upon the Weald Moors.** At Wappenshall junction (turn left where A442 meets B4394 and then take the first left again) warehouses and a toll collector's house can be seen where the Newport branch meets the Shrewsbury, although part of the latter has been filled in to make a private garden. From here you can follow the canal across fields to the next road. There are two locks in **Hadley Park,** and another on factory premises in Hadley on the north side of A518. On the south side of this road, two miles from its junction with A5, by a signpost pointing to the Shropshire Arms, are the remains of another lock and the approach to the foot of the **Trench incline,** alongside a reservoir that once fed the canal. The pub is by the bottom of the slope, 223 yards long with a rise of 75 feet. It was a double-track incline operated by a steam-engine, but there is little left at the top except a cottage and part of the entrance to the upper basin.

If you return to A518 and continue towards Newport you come to **Lilleshall.** In the fields behind the village there are traces of the Donnington Wood Canal. At **Pitchcroft** you may find the site of the terminal basin of the branch. A lane heading east from Lilleshall leads to **Hugh's Bridge,** where a sign to Incline Cottages gives a useful clue. A track to the cottages runs parallel to the canal, and the line of the incline can be seen over a farm gate and behind the cottages themselves. From here the main line continued to Pave Lane, 1½ miles south-east of Newport on A41. In the other direction it skirted the grounds of Lilleshall Abbey and can be followed from here to Muxton Bridge. It may not be possible to find much more of it.

For the Shropshire Canal, look for the **Wrockwardine Wood incline,** by the Bellevue Inn, 1¼ miles from Oakengates centre. By the junction of Furnace Lane, Plough Road and Moss Road a stony track rises, part of the incline, the rest of which has been swallowed up by development. From the top you can follow the canal for a short distance. This canal had two tunnels, at Snedshill and Stirchley, but both have gone. Between these tunnels was the junction with the Ketley Canal, most of which has disappeared. The top of the **Ketley incline** was near Ketley Hall and the foot near a pub, the Wren's Nest, in a housing estate south of A5 (turn southwards at a crossroads by the Seven Stars). Housing developments have destroyed most of the evidence.

The old Coalport branch railway took generally the direction of the Shropshire Canal, passing east of Dawley. On A442 heading south from Dawley is a village — a very small one — called **Aqueduct.** The aqueduct is on a dead-end road next to the main road; this carried the branch to Brierly Hill. You may be able to find a stretch of the branch above **Coalbrookdale,** on the west side of B4373. On this length there are two cottages said to have been designed by Telford, close to the top of the old railway incline.

The main line continued through **Windmill Farm,** where there was an inclined plane with a rise of 126 feet. However, this has been virtually obliterated by recent roadworks. It is best to continue through Madeley and make for the **Blists Hill Open Air Museum.** In its grounds is the last section of the Shropshire Canal, restored 1972-3; an ice-breaker and a restored iron tub-boat are afloat and a replica wooden tub-boat lies on the bank. The canal ends at the top of the **Hay incline;** the tub-boats were floated on to wheeled cradles and hauled up a short reverse slope by steam power to descend 207 feet to the basin at the foot. An empty boat ascending could be drawn up by a loaded boat on the way down. Rails have been relaid on the thousand-foot long slope and at the top are substantial remains of the docks and engine-house. The operation of the incline is fully explained in the Ironbridge Gorge Museum Trusts's pamphlet. At the foot of the incline, part of the half-mile-long **Coalport Canal,** parallel to the Severn, has been restored and leads you into the Coalport china works museum.

This is an area of inexhaustible fascination for anyone interested in industrial archaeology. In addition to what has been mentioned, you can see the Tar Tunnel — another undertaking of William Reynolds — the Blists Hill blast furnaces and mine, a sawmill, printing shop and a host of other reminders of the industrial past. Close by are the famous iron bridge over the Severn and the Coalbrookdale ironworks with its own museum. It all helps to put the canal firmly into the context of the age in which, and for which, it was built.

OS sheets 126 and 127 (1:50,000). 118 and 119 (1 inch).
Train: Shrewsbury and Oakengates.

13. The Somersetshire Coal Canal

Act: 1794. Opened: 1805. Abandoned: 1904. Length: 17¾ miles.

History

It takes an effort to realise that Somerset, rural as it is, was for centuries a coal-mining county. The very name of the Somersetshire Coal Canal reminds us, however, and the Act of 1794 'for making and maintaining a Navigable Canal with certain Rail Ways, and Stone Roads, from several Collieries, in the

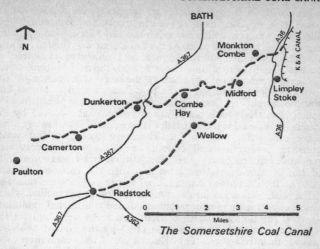

The Somersetshire Coal Canal

County of Somerset, to communicate with the intended Kennet and Avon Canal' makes the intentions of its promoters perfectly clear. There were to be two lines, one from Paulton and the other from Radstock, joining at Midford and thence connecting with the Kennet & Avon near Limpley Stoke. The Paulton line was completed and some part of the Radstock line; the latter was very little used and a railroad (horse-drawn) was laid along its course from Radstock to Twinhoe, where it connected with another length of railroad to Midford. At Midford, only a very short stretch of canal had been cut towards Radstock; it had ended at the foot of Midford Hill where a railroad had been constructed to overcome the problem of the steep gradient in the early years of the project.

The line of the SCC was first surveyed by John Rennie. Unlike most of the south-western canals, it was planned to take the conventional narrow-boats. William Smith, the geologist, took over as surveyor and worked for the company until 1799. He, with other committee members, toured England to examine the construction of canals and railroads elsewhere before work began. By 1797, good progress on the Paulton line was being made. There was, however, one major problem; the steep gradient at Combe Hay down which the canal had to descend to bring it to the level of the Cam and Midford brooks, parallel to which it was to flow to the K & A. Attempts to solve this problem took a great deal of money and delayed the opening of the canal for several years.

The first attempt was by means of a 'caisson lock', designed by

Robert Weldon, who demonstrated a prototype on the Shropshire Canal in 1794. This involved the construction of a masonry well, filled with water, in which a wooden box large enough to contain a narrow-boat was suspended. At the top a boat was floated into the box, which was then lowered by racks and pinions to the bottom of the well. There were sliding doors at both top and bottom, which made close contact with the ends of the box — the caisson. At the foot, the boat floated out of the box through a short tunnel and into the canal. The Combe Hay structure was completed in 1798 and several trials took place, one of which was watched by the Prince of Wales. He did not risk his own life in it but many other brave souls did, and no fatalities were recorded. But water pressure inside the well caused damage, which necessitated expensive repairs. The engineer Benjamin Outram was called in to advise; he recommended dispensing with the caisson lock and using an inclined plane instead. Coal was to be loaded in boxes on to the boats, unloaded into wagons at the top of the incline, which was double-track and counterbalanced, and transferred again into boats at the bottom. Three conventional locks were built at the foot of the incline. This was operating by the end of 1801 but proved unsatisfactory. The Kennet & Avon committee was becoming critical of the delays in completing the SCC and pressed for the installation of ordinary locks. A special Lock Fund was set up to raise extra money. By April 1805 the Combe Hay flight of twenty-two locks (including the three at the foot of the incline) was completed, with a total rise of 154 feet. Within a year a pumping engine was installed near the head of the flight to raise water to supply the locks. With an aqueduct at Midford, another pumping engine at Dunkerton and a lock at the junction with the K & A, the main line of the canal was now complete. The committee could not afford to finish the Radstock arm as a canal (in addition to the gradient problem, there was also a shortage of water) and agreed to a suggestion of a railroad along its line. This was completed in 1815. Further railroads — or tramroads — connected the various collieries with the canal at different places.

Including the Lock Fund, the SCC cost about £180,000 to construct. By 1820 it was carrying over 100,000 tons a year of coal from the twenty-three collieries it served. This rose to over 165,000 tons in 1858, but by this time tolls had to be cut to face railway competition and receipts were falling. Dividends remained good, for the company seemed to plough little money back into improving the canal itself. The K & A and the Wilts & Berks Canal, which carried on further the traffic originating on the SCC, were also suffering declining trade. When the Somerset & Dorset Railway proposed a line from Limpley Stoke to Radstock, the SCC realised the competition would be too strong and in 1871 sold their Radstock railroad to the railway company for £20,000. Railways were distributing cheaper coal from other areas, and the Somerset pits were closing down. The tonnage carried fell away sharply, dropping by 1890 to less than 20,000 a year. Now dividends ceased to be paid, expenditure exceeded income and the company went into liquidation in 1893. The canal was put up for auction but failed to reach the reserve of £3,900. A little trade continued until 1898; then the canal was closed. In 1904 it was sold to the GWR who built the Camerton to Limpley Stoke railway along part of its course. This line was closed in 1951. The following year it was used for the film *The Titfield Thunderbolt*, the track not being lifted until 1958. The Somerset & Dorset line was closed in 1966.

The canal today

Most canal enthusiasts know Rennie's superb **Dundas aqueduct,** which carries the K & A over the Avon and the railway line. It is on the A36 Bath-Warminster road, about 1½ miles south of Claverton. At the west end of the aqueduct is the junction with the SCC; the top of the lock, now a flower bed, can be seen in the garden of a cottage which was the lock-keeper's cottage. The **Viaduct Hotel** marks where the road crossed the canal. The course, used also by the railway, runs parallel to the Midford Brook and south of **Monkton Combe.** The minor road from A36 through Monkton Combe leads past a cottage which belonged to William Smith, between Monkton Combe and **Midford.** At Midford there was a weigh-house, long since demolished; the aqueduct that took the Radstock arm over the brook is still there, but crumbling away.

Continue along the minor road towards Combe Hay. A railway bridge crosses the entrance to a field on your right, in about a mile. In the field through this bridge are the remains of many of the **Combe Hay** locks. You can follow them around the outside of the field in a semi-circle. They change direction to head west, alongside a lane leading to Caisson House; you should obtain permission to explore further. The locks cross in front of the house to the top level of the canal on its west side. The inclined plane descended by the fifth lock from the top. The pumping house was on the east side of Caisson House in Engine Wood, joined to the canal by a short arm. Attempts to discover the exact site of the caisson lock have so far been unsuccessful.

On the west side of Combe Hay village, partly concealed by rubbish tips, is a short tunnel used first by the canal and then by the railway. Continue towards **Dunkerton;** look for remains of a canal bridge in a lay-by just north of where the minor road crosses A367. You can also see an earth aqueduct that took the canal across a valley. Go on to **Camerton;** you will see the canal line from the top of the tip and behind the Jolly Collier. The canal passed under the road at Radford and can be followed from here to the basin at **Paulton,** almost due west.

Of the Radstock branch there is very little to be seen. The basin at **Radstock,** south of the Waldegrave Arms, has been obliterated. Near **Shoscombe** there is a small aqueduct and at **Wellow,** in a farmyard on the east side of the church, is the south portal of a 405-foot tunnel which took the canal and then the railroad which soon replaced it. Part of the tunnel is now an elongated farm shed. It may be possible to find traces of the top of the north portal in a field behind the church.

OS sheet 172 and 183 (1:50,000), 166 (1 inch).

14. The Thames & Severn Canal

Act: 1783. Opened throughout: 1789. Cirencester Branch opened c 1787. Abandoned: Inglesham — Chalford 1927; Chalford — Stroud 1933. Length 28¾ miles. (Cirencester Branch 1½ miles).

History

The Thames & Severn Canal was the major element in an ambitious scheme to link the most important trading rivers in southern England. It was a subject for discussion for nearly two hundred years before it was eventually opened. Words were not translated into action until the opening of the Stroudwater Canal in 1779, linking the busy wool town of Stroud with the Severn 8 miles away. Stroudwater representatives joined with those of some of the Midland canal companies and with a number of London merchants to have a line surveyed by Robert Witworth. Agreement was soon reached and the first steps were taken to raise funds. Many of the original shareholders were also proprietors of the Staffordshire & Worcestershire Canal and it was one of them, James Perry, who became first superintendent of works on the Thames & Severn. Josiah Clowes was appointed engineer, and six years after the Act was obtained the canal was open from Stroud to Lechlade.

The Thames & Severn was built as a broad canal with forty-four locks. From Stroud to Brimscombe, which was developed into a large inland port, the locks were made to the same dimensions as those on the Stroudwater — 72 feet long and 15 feet 6 inches wide, able to take the Severn trows. The locks east of Brimscombe were longer but too narrow for the trows; hence goods were transhipped at the port from trows into barges or narrowboats, and vice versa. Twenty-eight locks lifted the canal 241 feet from Stroud to the summit level. Here the Cotswold escarpment was pierced by the great Sapperton tunnel, the third longest canal tunnel to be built in Britain. Sixteen more locks brought the canal down to the junction with the Thames at Inglesham on the outskirts of Lechlade. There were no locks on the short branch to Cirencester.

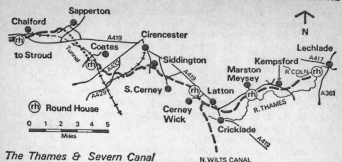

The Thames & Severn Canal

But the hopes of the promoters were not fulfilled. Two other cross-country canals, the Kennet & Avon and the Wilts & Berks, both opened in 1810 and provided competition. The navigation of the upper Thames was in poor condition and deterred traders from using the canal as a through route to London. There were recurring difficulties over water supply at the summit, as well as leakage in places and trouble with the tunnel. In the mid nineteenth century the Great Western Railway made damaging inroads into the canal's trade. In 1882 the GWR took over control of the canal mainly, it seems, in order to prevent anyone else trying to develop it in any way. An attempt at revival was made by a number of canal companies and local authorities who formed a trust to run the canal in 1895. This scheme failed, and six years later the trustees handed the canal over to Gloucestershire County Council. Despite repairs and improvements, the traffic, once lost, never returned. The last through voyage was made in 1911, and the last stretch to survive, from Chalford to Stroud, was abandoned in 1933.

The canal today

From **Stroud** to Sapperton tunnel the line of the canal is easy to follow. The Thames & Severn met the Stroudwater at **Wallbridge**, on A46 a few hundred yards south of the railway bridge. From here you can walk the towpath, with only a few short diversions, as far as the tunnel, about 8½ miles. The first diversion comes almost immediately; after inspecting the remains of the basin and wharf you have to go to the A46/A419 junction where you can rejoin the canal by an iron bridge to the right. The Stroudwater, Thames & Severn Canal Trust has repaired a lock and some bridges on this section, part of which has also been dredged. The Ship Inn at **Brimscombe** used to be a canalside pub; a few yards to the east Benson's factory occupies the site of the old inland port. Here was the centre of the canal company's trading operations; there was a basin 700 feet long, sheds, houses, a coal store and a large, well-proportioned warehouse. Now there are only two of the smaller buildings, one a wharfinger's cottage and the other a salt store, remaining by the entrance to the factory yard. Two boat-building firms had their premises by the canal near Brimscombe.

The towpath can be regained along a lane between the river Frome. and the factory and can be followed as far as Chalford. Between Stroud and Chalford the canal is throughout close to the south side of A419 and there are plenty of easy access points to it from the main road. At **Chalford**, just before this road and the canal part company, is a round house, one of the five built on the T & S to accommodate the lock-keepers or 'watchmen'. Within a few hundred yards the main road crosses the canal; the railway, which has hitherto kept both company, continues to follow the canal line for a couple more miles. The large and attractive village of Chalford struggles up the hillside. To follow the canal by road you have either to take the lower road through Chalford, turn up over the hills to Oakbridge and then back down towards Sapperton, or to return to the main road, turn left for Frampton Mansell and head for Sapperton on the south side of the canal. Best of all is to walk there; there are ten locks up to the summit and the countryside is beautiful, the hills rising steeply on either side. Whichever way you go, your target is the **Daneway Arms,** which used to be called the Bricklayers' Arms and was built in 1784. The pub car park covers the site of the top lock; from here it is a short walk to the west portal of the tunnel.

Sapperton tunnel is now blocked by roof falls. The west portal, resembling the dilapidated entrance to a medieval castle, may soon be restored. A419 crosses the line of the tunnel on the north side of Hailey Wood; it is easily distinguished by the clumps of trees planted on the spoil heaps to improve the landscape in Earl Bathurst's park. The road also crosses the line of the shorter railway tunnel. The east portal of Sapperton is reached off the minor road between Coates and Tarlton. This road crosses the canal; a signpost directs you along a lane parallel to a good firm stretch of canal bed in a wooded cutting, holding water. This was one of the stretches improved in the ownership of the County Council. The lane crosses above the portal and stops at the **Tunnel House,** which once provided lodgings for the tunnel diggers and then for the boat crews, but is now an inn. The classical portal has been beautifully restored by the Canal Trust; the towpath has been cleared and the bed of the canal cleaned out.

On the other side of the Coates-Tarlton road you can follow the towpath to **Coates** round house, roofless and derelict. Unless you are walking the whole length, it is best to return to the road and turn back towards Coates. Turn right before you reach the village and continue to A433. Turn right again; in a few hundred yards there is a lay-by and the parapets of **Thames Head** bridge — note the plaque but watch the traffic! To the west is the supposed source of the Thames, though the statue marking the spot has been moved to St John's lock. The bridge has been recently bypassed by a culvert. The towpath here is not a public right of way, but the line, continuing eastwards, soon reaches a group of buildings. This is the site of **Thames Head Pumping Station.** The canal company soon found that there was a shortage of water at the summit and installed first a windmill pump and then a steam-engine. This was replaced by a Cornish beam engine in 1854, capable of raising 3 million gallons of water every twenty-four hours. Only a dwelling house and a few outbuildings are left.

The dry bed of the canal straggles across the fields towards Siddington. Once it crossed the Kemble-Cirencester road by the **Smerril aqueduct,** of which the abutments can be seen two miles from Kemble. At **Siddington** there are the remains of a lock and the junction with the Cirencester branch on the west side of the village; of the branch itself hardly anything is left. Siddington is the end of the Summit level with four locks close together each with a fall of 9 feet 9 inches. A house has been built on the site of the bottom lock. From Siddington to the top of the three **South Cerney** locks the towpath is a right of way; the locks have been filled in but the coping stones of the top one are visible in the garden of the former lock cottage. The canal is intact, except for three culverted road crossings, from the bottom South Cerney lock to Latton. To explore this stretch, take the Spine Road off A419 at a length of dual carriageway between Cirencester and Cricklade and park at the tourist information lay-by, which is beside the canal. Northwards you can find the two Wildmoorway locks and southwards Cerney Wick lock and round house.

Continuing towards Cricklade on A419 you come to the out-skirts of **Latton.** A lane by the group of houses on your right leads to the junction of the Thames & Severn with the North Wilts Canal; by this route boats could reach the Wilts & Berks Canal at Swindon. Onwards from Latton you can see the **Cricklade** wharf house from the bypass opposite the turning to Kempsford. Follow the Kempsford road; the course of the canal runs through the fields on your right. There is a left turn to Marston Meysey: a track on the right just past this leads to **Marston Meysey** round house, empty and disused but rendered in order to preserve it. At **Kempsford** there is another wharf house.

From this point access by road is difficult as the line lies between the Thames and the Coln. It is necessary to make for **Lechlade.** Then you can either walk alongside the Thames upstream for about half a mile to see the junction of the canal with the river by the **Inglesham** round house or, if you wish to get closer, take A417 westward from Lechlade and turn down a lane to the left, half a mile from the village centre. The round house is private property, but you can see the lock chamber and bridge.

The Stroudwater, Thames & Severn Canal Trust has been involved with work at three sites on the Thames & Severn and at Eastington on the Stroudwater and there is hope that, in time, these canals will be fully restored.

OS sheets 162 and 163 (1:50,000), 156 and 157 (1 inch).
Train: Stroud.

15. The Wey & Arun Junction Canal

Act: 1813. Opened: 1816. Abandoned: 1868. Length 18½ miles.

History

The river Wey, from the Thames to Guildford, was navigable in the 1660s and the navigation was extended to Godalming in 1763. In 1794 the Basingstoke Canal was opened, adding to the trade on the river which was making healthy profits mainly from carrying coal, timber and flour. The river Arun was made navigable towards the end of the eighteenth century. It seemed logical to make a link between the two rivers, so that there could be a through navigation from London to the Sussex coast — even, indeed, to Portsmouth by means of another connecting waterway. With the support of the wealthy Lord Egremont, an Act was obtained in 1813 for a canal to join the Wey and Arun at an estimated cost of rather less than £100,000. The line did not take long to complete, being opened with twenty-three barge locks and eight wharves three years later. An Act was passed in 1817 for a canal joining the Arun at Ford, near Arundel, to Portsmouth; this was opened in 1823, with a branch to Chichester, at a cost of £170,000.

It was an ambitious scheme which never met the success it deserved. The Arun proved an unsatisfactory navigation and the capacity of the barges was too small to interest traders wishing to use the whole length of the route. When the wars with France ended, coastal trade between London and Portsmouth revived. The Portsmouth — Arundel line attracted little traffic, being slow and difficult as much of it was through tidal water. So the through traffic anticipated did not materialise. Tonnage carried annually on the Wey & Arun generally remained below 20,000, exceeding that figure in only seven years, and the highest figure for receipts was a mere £2,524, in 1839. Railways were late to come to this area, but when the Horsham & Guildford line was opened in 1865 it became clear to most of the canal's shareholders that there was no commercial future for their undertaking. An Act to abandon the canal was passed in 1868 and the canal was closed three years later.

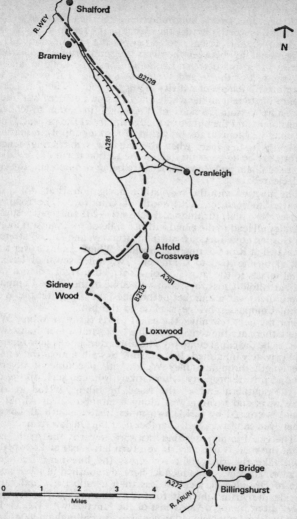

The Wey & Arun Junction Canal

The canal today

At this point, over a hundred years after the closure of the Wey & Arun, it is worth quoting the object of the recently formed Wey & Arun Canal Trust: 'to attempt the restoration of the navigational link between the rivers Wey and Arun, thus providing a direct water link between the South Coast and London, and also with the rest of the Inland Waterway system.' The explorer may therefore find signs of activity along the route. The Trust is not attempting to reopen the whole line by its own efforts, but intends to restore certain sections of the canal in order to convince government and the public of the feasibility of the project. One of the chief problems is the provision of a water supply, a matter of difficulty in the years when the canal was a trading concern. Efforts are being concentrated on the section north of Newbridge, where Malham and Rowner locks are being restored and the canal bed cleared.

The junction with the Wey, still a navigation, is at Stonebridge Wharf, **Shalford,** on A281 south of Guildford. The road soon crosses the canal; turnings eastward off A281 for 6 miles south of Bramley all lead to the canal within a mile or so. Points to look for are **Gosden aqueduct,** just north of Bramley, and a good stretch of canal bed at **Run Common. Elmbridge Wharf** is a little over a mile east along the road to Cranleigh. Two miles south of here, the canal moves to the west side of A281. Turn right on to a minor road at Alfold Crossways; soon you come to the **Three Compasses,** where there was a banquet before the first voyage on the Wey & Arun. Compasses bridge is close to the pub.

For the next few miles the canal winds through Sidney Wood, where there are the remains of nine locks and a lock house which used to be the canal company's headquarters and is now renovated and privately inhabited. It is possible to walk almost the whole of the length through Sidney Wood, with a couple of diversions around private property. Alternatively, you can pick up the canal at the southern end of the wood by taking B2133 at Alfold Crossways and turning west along a minor road at Alfold. The canal is crossed by B2133 two miles further south at **Loxwood,** where you can find a lock chamber near the Onslow Arms.

The canal is now heading eastward towards the Arun; access from the road is not easy. If you turn left (east) at Loxwood and then take the first right you cross the line near Drungewick Manor. There was an aqueduct here, demolished in 1957, and the site of **Drungewick lock.** In less than a mile the canal runs up beside the Arun which it follows southward to **Newbridge,** on A272 a few hundred yards east of the junction with B2133. From here on, the Arun Navigation continues to Pallingham lock.

There are remains of the Portsmouth-Arundel canal to be found at Ford, by the Ship and Anchor, and at Yapton, on A2024, behind the Shoulder of Mutton and Cucumbers. The Chichester branch still exists south of the city; look for Basin Road. The Portsmouth end of this waterway was formed by the Portsea Canal; most of this has disappeared under roads or railway development, but you can find a surviving stretch by the headquarters of the Langstone Harbour Fishermen's Association.

A first-class historical account of the London-Portsmouth waterway line can be found in *London's Lost Route to the Sea*, by P. A. L. Vine. Information can also be obtained from the Wey & Arun Canal Trust.

OS sheets 186 and 197 (1:50,000), 182 and 170 (1 inch).
Train: Guildford, Shalford, Billinghurst (then) bus to Newbridge).

16. The Wilts & Berks Canal

Act: 1796. Opened throughout: 1810. Abandoned:1914. Length: 51 miles, with about 6½ miles of branches. North Wilts Canal: 9 miles.

History

The Wilts & Berks was the third of the major east-west water routes in the southern half of England. It was cut from the Kennet & Avon, at Semington, near Melksham, to Abingdon on the Thames. By means of the North Wilts Canal, opened in 1819, it connected with the Thames & Severn at Latton, near Cricklade. Branches were built to Calne, Chippenham, Longcot and Wantage. Its forty-five locks were built to narrowboat dimensions, unlike those of the K & A and T & S; nor did it have any major engineering works to rival the Dundas aqueduct or Sapperton tunnel. It was a long, meandering waterway; there were various proposals for expansion but generally they came to nothing, and its history is comparatively uneventful. Coal was the main traffic; agricultural produce was also distributed along its line. Most of the coal came from the Somerset mines, via the Somersetshire Coal Canal and the K &A; most of the loaded boats, therefore, travelled eastwards as there was little return traffic. Tonnage carried was never very great, rising to over 60,000 in the late 1830s and early 1840s; dividends rose to £9,000 in the most profitable years. Trade in these years, however, was augmented by the carriage of material for the construction of the Great Western Railway's line through Swindon and Chippenham; when this was completed, the canal's prosperity rapidly diminished. As receipts fell away so the standard of maintenance declined. Despite economies, by the 1870s the canal was running at a loss. The canal company tried, but failed, to sell out to the GWR, and then determined on closure or sale to some other body. In 1875 it was bought for £13,466 by a group of merchants who traded upon it;

they ran it for a few years and then leased it to some Bristol traders for £1,250 *per annum*. The Bristol group battled on for another five years but gave up in 1887. Its owners struggled with it themselves until 1891 when a new company, called the United Commercial Syndicate, was formed to take it over. They put some money into it and ran a regular fly-boat service, but they too were unable to make a profit.

By 1897 the end was in sight. The owners now wanted to abandon the canal, and so did the Swindon Traders Association who considered it unnecessary and a nuisance. But it was not until 1914 that an Abandonment Act was obtained, partly because local landowners wanted it kept in being for water supply. By 1906 commercial traffic had ceased; in the following years parts became stagnant and others dry, several bridges were in poor condition and many locks unusable. On abandonment, Swindon took over the reservoir at Coate for water supply and much of the canal was given to those through whose land it ran. Lady Wantage, whose husband had provided much of the money for the United Commercial Syndicate, was owed £16,000 when the canal was closed, but received nothing. The Wilts & Berks had cost over £250,000 to build, but had never come near to realising the expectations of its promoters.

The canal today

Much of the Wilts & Berks can still be traced, and some stretches provide drainage for farmers' fields. L. J. Dalby's study of the canal, referred to in the bibliography, provides a detailed itinerary for anyone wishing to explore the whole length. Here I can mention some of the points easily accessible from roads; the connecting sections can be seen on the relevant OS maps.

The Wilts & Berks left the K & A on the east side of **Semington** bridge, on A350 about two miles south of the centre of Melksham. The entrance to the K & A, under a bridge carrying the towpath, has been filled in; the W & B lock is in the garden of what was the toll collector's house, still inhabited. There are traces of the canal on the east side of the road into **Melksham;** a line of trees and a footpath give some clues and there is a bridge parapet on the corner of Forest Road. From Melksham the canal heads towards Chippenham, keeping on the east side of the Avon. There is a small aqueduct over Forest Brook, behind Forest Farm, and the remains of a couple of locks on this stretch. The canal passes east of **Lacock** and the line is crossed by the two minor roads running east from this most attractive and beautifully preserved village. The Chippenham branch left the main line near another Forest Farm, to the south of the A4 Chippenham-Devizes road. In **Chippenham,** a bus station has obliterated the basin and the short tunnel leading to the basin has disappeared.

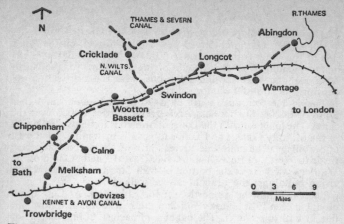

The Wilts & Berks Canal

The main line continues on the east side of A4. It is worth making for **Studley** and taking the minor road leading north-west from the village. This crosses the canal just past Studley Abbey Farm. You can follow it by the site of a bridge, which once carried the Calne-Chippenham railway, and carry on to the two-arched aqueduct over the Marden. One arch has partly collapsed — this happened in 1906 — but it may still be possible to inch one's way across it with care. Soon you will come to the remains of the two **Stanley** locks and then to the junction with the branch to Calne. Much of this branch survives, and the wharf in **Calne** is used as a car park by Harris's.

The main line continues north-north-west, being crossed by minor roads near East Tytherton and Foxham. It draws nearer to the line of its successful competitor, the railway, almost coinciding with it at **Dauntsey Lock**, where the railway and canal are crossed by A420. You can see the site of the wharf, and the lock house is inhabited. There were seven locks between here and Wootton Bassett; you can pick up the canal from the three minor roads heading northwards off A420 between Lyneham and the next railway bridge.

The canal, now heading eastward, passes south of **Wootton Bassett** and is crossed by minor roads southwards off A420 to the outskirts of Swindon. In **Swindon** the line of the canal is a grassed-over footpath leading towards the Brunel shopping centre, passing under Cambria and Milton Road bridges. Rope grooves can be seen on the iron stanchions of Cambria bridge. The old Wharf Road is now Market Street. Off Cromwell Road a car park has been constructed on the line and the Corporation Wharf is now grassed. In the shopping centre the line is indicated by Canal Walk; a milestone is preserved near Regent Street, on the other side of the canal from its former position. The Parade takes the line to Fleming Way and the site of the junction with the North Wilt. Fleming Way continues the line, with the fire station occupying the site of Swindon wharf.

The course continues south of A420 to **Shrivenham** and then south of B4508. Shrivenham Arch bridge carries B4000 over the canal. The junction with the Longcot branch is in fields a mile east of the Military College and **Longcot** basin is by a minor road west of the village. The main line curves north of Uffington and across fields to West and East Challow. B4001, A417 and A338 cross it. **Wantage** wharf is at the bottom of Mill Street, with several surviving buildings; the branch can be walked.

The canal now heads north-east for Abingdon. At **Cow Common,** between East Hanney and Steventon, a footpath follows the line in both directions. The canal approaches **Abingdon** on the south side of the Ock. It circles a housing estate by A34 and takes the course of Caldecott Road. You can find the bridge over the Ock, erected in 1824. South of the bridge by Wilsham Road the canal entered the Thames.

The longest branch, the North Wilts Canal, leaves the main canal in Fleming Way, **Swindon,** becoming a path under the railway works and out towards Rodbourne Road. A lock chamber was uncovered near the gasometer at Iffley Road. North-west of Swindon, head for a minor road running due north from Common Platt to Cricklade. About a mile north of Common Platt the canal and road virtually coincide; watch the east side of the road for evidence. Where this road meets B4041 there is an aqueduct over the Key. The canal continues to the west of Cricklade; search for the southern portal of a 200-yard tunnel and the remains of an aqueduct over the Churn.

In 1977 the Wilts and Berks Canal Amenity Group was formed, with the aim of protecting the line of the canal and its branches from further destruction. Current projects include clearing a stretch at Kingshill, Swindon, to form a water park, removing rubbish from under Shrivenham Arch bridge and surveying the buildings by Dauntsey Lock.

OS sheets 173, 174, 164 (1:50,000), 166, 156, 157, 158 (1 inch).
Train: Chippenham, Swindon.

17. Some other lost canals

In addition to the canals described in this book, there are many others with remains that are well worth exploring. The following list includes some of the more interesting ones, with their lengths and terminal points.

South-west England

Glastonbury Canal: Highbridge — Glastonbury (14⅛ miles).

Liskeard & Looe Union Canal: Moorswater, near Liskeard — Terras Pill, near East Looe (5⅞ miles).

St Columb Canal: Trenance Point — St Columb Porth, but central section not completed (about 6 miles).

Stover Canal: river Teign, near Newton Abbot — Teigngrace (1⅞ miles).

Tavistock Canal: Tavistock — Morwellham (4 miles, plus branch to Millhill, 2 miles).

Torrington Canal: Torrington — river Torridge near Bideford (6 miles).

South and South-east England

Andover Canal: Andover — Redbridge (22 miles).

Royal Military Canal: Shorncliffe — Winchelsea (30 miles).

Salisbury & Southampton: Alderbury Common — Kimbridge, and Redbridge — Southampton (total 13 miles).

Thames & Medway Canal: Gravesend — Frindsbury (7 miles).

Midlands

Some parts of the Birmingham Canal Navigations.

Buckingham Branch of the Grand Union Canal: Old Stratford — Buckingham (10½ miles).

Cromford Canal: Cromford — Langley Mill (14⅛ miles).

Nottingham Canal: Nottingham — Langley Mill (14¾ miles).

Oakham Canal: Oakham — Melton Mowbray (15¼ miles).

Oxford Canal: some short loops cut off when the canal was straightened.

Uttoxeter Branch of the Caldon Canal: Froghall — Uttoxeter (13¼ miles).

Eastern England

Horncastle Canal: Horncastle — river Witham (11 miles).

Louth Canal: Louth — Tetney Haven (12 miles).

North Walsham & Dilham Canal: Wayford bridge — Antingham ponds (8½ miles).

North-east England

Barnsley Canal: Barnby Basin — Heath, on Aire & Calder (15 miles).

Driffield Canal: top 5 miles of Driffield Navigation.

Leven Canal: Leven — Hull river (3¼ miles).

Market Weighton Canal: Weighton lock on the Humber — Weighton Common (9 miles).

North-west England

Carlisle Canal: Carlisle — Port Carlisle (11¼ miles).

Manchester, Bolton & Bury Canal: river Irwell, Manchester — Bolton, with branch to Bury (total 15¾ miles).

St Helen's Canal (Sankey Brook Navigation): Sankey Bridges, near Warrington — Blackbrook (8 miles, with some short branches).

Ulverston Canal: Ulverston — estuary of river Leven (1½ miles).

South Wales

Glamorganshire Canal: Merthyr Tydfil — Cardiff (25½ miles).

Monmouthshire Canal: Newport — Pontymoile (joins Brecon & Abergavenny Canal), also branch from Crumlin joining main line at Malpas (main line 11 miles, branch 11 miles).

Neath Canal: Glyn-Neath — Giant's Grave, near Neath (13 miles).

Swansea Canal: Swansea — Abercrave (15½ miles).

Tennant Canal: Port Tennant — junction with Neath Canal at Aberdulais (4⅞ miles).

---

18. Lost canals found

Canals on which restoration work is actively proceeding and which are navigable in part include the following:

Basingstoke Canal, from Basingstoke to the river Wey near Weybridge. About half of the restoration programme is completed.

Droitwich Canal, from the Severn to Droitwich and thence by the Droitwich Junction Canal to the Worcester & Birmingham Canal. Restoration is in full progress.

Kennet & Avon Canal, from Bath to Newbury. Many miles are now navigable and work is beginning on the flight of twenty-nine locks near Devizes.

Montgomeryshire Canal, from Newtown to Frankton. Restored from Welshpool to Burgedin Lock, near Arddleen. Stretches either side of the Vyrnwy aqueduct restored and aqueduct repaired.

Pocklington Canal, from Pocklington to East Cottingwith on the Derwent. 4½ miles at the western end now restored.

Further reading

Historical studies of all the canals mentioned in this book can be found in 'The Canals of the British Isles' series, published by David & Charles (Newton Abbot). This series comprises the following volumes:

British Canals, Charles Hadfield (5th edition, 1974). Introductory volume to the series.
The Canals of the East Midlands; Charles Hadfield.
The Canals of North West England; Charles Hadfield and Gordon Biddle.
The Canals of South and South East England; Charles Hadfield.
The Canals of South Wales and the Border; Charles Hadfield.
The Canals of South West England; Charles Hadfield.
The Canals of the West Midlands; Charles Hadfield.
The Canals of Yorkshire and North East England; Charles Hadfield.
The Canals of Eastern England; John Boyes and Ronald Russell.

Lost Canals of England and Wales, by Ronald Russell (David & Charles, 1971), gives brief accounts of eighty navigations.
Britain's Lost Waterways; M. E. Ware; Moorland Publications, 1979. A pictorial record.

Studies of individual canals include the following:
The Somersetshire Coal Canal and Railways; K. R. Clew; David & Charles, 1970.
The Thames & Severn Canal; H. Household; David & Charles, 1969.
The Dorset & Somerset Canal; K. R. Clew; David & Charles, 1971.
London's Lost Route to the Sea; P. A. L. Vine; David & Charles, 1965.
The Bude Canal; H. Harris and M. Ellis; David & Charles, 1972.
The Grand Western Canal; H. Harris; David & Charles, 1973.
The Wilts & Berks Canal; L. J. Dalby; Oakwood Press, 1971.
The Chard Canal; Chard History Group.

Useful general and reference books on canals include:
The Inland Waterways of England; L. T. C. Rolt; Allen & Unwin, 1950.
Navigable Waterways; L. T. C. Rolt; Longmans, 1969.
The Canal Age; Charles Hadfield; David and Charles, 1968.
The Canal Builders; Anthony Burton; Eyre Methuen, 1972.
The Shell Book of Inland Waterways; H. McKnight; David & Charles, 1975.
The Complete Book of Canal and River Navigations; E. W. Paget-Tomlinson; Waine Research Publications, 1978.
Historical Account of the Navigable Rivers, Canals and Railways throughout Great Britain; Joseph Priestley; 1831, reprinted David & Charles, 1969.
Bradshaw's Canals and Navigable Rivers; H. R. de Salis; 1904, reprinted David & Charles, 1969.

Index